Food Safety and Food Quality

ISSUES IN ENVIRONMENTAL SCIENCE AND TECHNOLOGY

TITLES IN THE SERIES:

How to obtain future titles on publication

A subscription is available for this series. This will bring delivery of each new volume immediately upon publication and also provide you with online access to each title via the Internet. For further information visit www.rsc.org/issues or write to:

Sales and Customer Care Department, Royal Society of Chemistry, Thomas Graham House, Science Park, Milton Road, Cambridge CB4 0WF, UK

Telephone: +44 (0) 1223 420066
Fax: +44 (0) 1223 423429
Email: sales@rsc.org

ISSUES IN ENVIRONMENTAL SCIENCE AND TECHNOLOGY

EDITORS: R. E. HESTER AND R. M. HARRISON

15

Food Safety and Food Quality

RS•C
ROYAL SOCIETY OF CHEMISTRY

ISBN 0-85404-270-9
ISSN 1350-7583

A catalogue record for this book is available from the British Library

Published by The Royal Society of Chemistry, Thomas Graham House,
Science Park, Milton Road, Cambridge CB4 0WF, UK
For further information see our web site at www.rsc.org

Typeset in Great Britain by Vision Typesetting, Manchester
Printed and bound by Bookcraft Ltd, UK

Preface

Among the popular concerns about the environment, none ranks higher than the safety and quality of the food that we eat. Issues relating to genetically modified (GM) crops, pesticide residues in fruit and vegetables, cancer-causing chemicals, hazardous micro-organisms such as *E. coli*, the incidence and control of transmissible spongiform encephalopathies such as BSE and CJD, are but a few of the problem areas that receive frequent and widespread coverage in the media. In this book we have brought together a group of experts to present an up-to-date and balanced overview of a wide range of these issues, providing authoritative detail in a readily accessible way.

Andrew Chesson of the Rowett Research Institute in Aberdeen has provided an assessment of the safety of GM food crops which addresses this controversial subject rationally. His account gives careful consideration to the identification of possible hazards, however remote or implausible, but also to the analysis of risk. He concludes that although the present generation of GM crops may pose some risks to the wider environment, there is, in fact, no evidence that suggests they pose any risk to human health.

On a broader front, the views of the UK food industry on safety and quality research priorities are represented in an article by Geoff Andrews of Northern Foods, Alistair Penman of Unilever and Chris Hart of Weetabix Ltd. They review a wide range of microbiological agents known to be involved in food poisoning incidents (*e.g. Campylobacter* spp., *Cryptosporidium*, *Salmonella* spp., *E. coli* 0157) and discuss the use of Hazard Analysis and Critical Control Points (HACCP) as a risk management tool. Chemical safety issues also are addressed, such as those associated with veterinary residues (hormones, antibiotics), GMOs, dioxins from incinerators, and marketing considerations such as 'shelf-life'.

A more detailed examination of the food safety issues associated with the widespread and increasing practice of recycling biosolids (sewage sludge, agricultural wastes and exempt industrial wastes) to agricultural land is provided in the following article by Jim Wright, Technical Director of a major global consultancy organization, Environmental Resources Management. He compares the legislative framework and industry practice in the UK and USA, and provides a wealth of detail on the pathogens present in sewage sludge and other such biosolids, identifying the benefits of applying HACCP methods to the recycling process.

David Taylor, who currently runs the consulting service SEDECON2000, draws on his long experience in researching the nature, causes and treatment of transmissible degenerative encephalopathies (TDEs) in a vividly explicit review of the BSE/CJD problem which continues to excite much public debate and concern. Under the title 'Mad Cows, Demented Humans, and Food' he takes a serious and wholly authoritative look at the many aspects of this disease, from the prion proteins believed to be essential components of the infectious agents to the practical measures introduced to enhance the safety of bovine-derived food products.

Next comes a stimulating and thought-provoking article on the natural and synthetic chemicals in the normal diet that may cause cancer. This is by Lois Gold, Thomas Slone and Bruce Ames of the University of California, Berkeley, and other collaborating institutions in California, USA. The article examines critically the assumptions, methodology, results and implications of regulatory cancer risk assessments of synthetic chemicals and compares these with naturally occurring chemicals in food. Extensive listings with hazard assessments for the chemical components of potatoes, coffee, bacon, bread and very many other common foodstuffs are given. Their critical evaluation of the available data leads the authors to conclude that standard methods for cancer risk assessment of synthetic chemicals are flawed and can greatly exaggerate risks, while, conversely, the risks associated with many natural chemicals that are found widely in food are commonly underestimated.

The current aim of the UK Ministry of Agriculture, Fisheries and Food (MAFF) is 'To ensure that consumers benefit from competitively priced food, produced to high standards of safety, environmental care and animal welfare and from a sustainable, efficient food chain; and to contribute to the well-being of rural and coastal communities'. Christina Goodacre, who is the Head of the Food Technology Unit, has provided an account of some of the research funded by MAFF through their Food LINK Programmes. These examples involve the areas of food flavour and polymer science as well as vision-based modelling of food quality.

Finally, from Peter Lillford of Unilever Research, we have a view of the assessment of food quality. This addresses the issues of quality control and quality assurance, underlines the important role of the consumer and expert tasting panels, and discusses such difficult concepts as perceived flavour and texture, odour perception and 'taste in the mouth'.

The book does not pretend to be fully comprehensive, but we believe it does provide an authoritative, in-depth but thoroughly readable account of a set of topics which are of central importance to both producers and consumers of food, *i.e.* to all of us. We anticipate that it will be of interest to a wide readership, to food scientists in industry, in research institutes, and in consultancy, to legislators and to government officials and advisors. Certainly it is expected to be essential reading for students in many food science and environmental science courses. We also commend it to the general reader, who will find much of interest in these articles.

Ronald E. Hester
Roy M. Harrison

Contents

Contents

Editors

Ronald E. Hester, BSc, DSc(London), PhD(Cornell), FRSC, CChem

Ronald E. Hester is Professor of Chemistry in the University of York. He was for short periods a research fellow in Cambridge and an assistant professor at Cornell before being appointed to a lectureship in chemistry in York in 1965. He has been a full professor in York since 1983. His more than 300 publications are mainly in the area of vibrational spectroscopy, latterly focusing on time-resolved studies of photoreaction intermediates and on biomolecular systems in solution. He is active in environmental chemistry and is a founder member and former chairman of the Environment Group of the Royal Society of Chemistry and editor of 'Industry and the Environment in Perspective' (RSC, 1983) and 'Understanding Our Environment' (RSC, 1986). As a member of the Council of the UK Science and Engineering Research Council and several of its sub-committees, panels and boards, he has been heavily involved in national science policy and administration. He was, from 1991 to 93, a member of the UK Department of the Environment Advisory Committee on Hazardous Substances and from 1995 to 2000 was a member of the Publications and Information Board of the Royal Society of Chemistry.

Roy M. Harrison, BSc, PhD, DSc (Birmingham), FRSC, CChem, FRMetS, FRSH

Roy M. Harrison is Queen Elizabeth II Birmingham Centenary Professor of Environmental Health in the University of Birmingham. He was previously Lecturer in Environmental Sciences at the University of Lancaster and Reader and Director of the Institute of Aerosol Science at the University of Essex. His more than 300 publications are mainly in the field of environmental chemistry, although his current work includes studies of human health impacts of atmospheric pollutants as well as research into the chemistry of pollution phenomena. He is a past Chairman of the Environment Group of the Royal Society of Chemistry for whom he has edited 'Pollution: Causes, Effects and Control' (RSC, 1983; Third Edition, 1996) and 'Understanding our Environment: An Introduction to Environmental Chemistry and Pollution' (RSC, Third Edition, 1999). He has a close interest in scientific and policy aspects of air pollution, having been Chairman of the Department of Environment Quality of Urban Air Review Group and the DETR Atmospheric Particles Expert Group as well as currently being a member of the DETR Expert Panel on Air Quality Standards and the Department of Health Committee on the Medical Effects of Air Pollutants.

Contributors

B. N. Ames, *Department of Molecular and Cell Biology, University of California, Berkeley, CA 94720, USA and Children's Hospital of Oakland Research Institute, Oakland, CA 94609, USA*

G. Andrews, *Northern Foods plc, Farnsworth House, Lenton Lane, Nottingham NG7 2NS*

A. Chesson, *Rowett Research Institute, Bucksburn, Aberdeen AB21 9SB*

C. Goodacre, *Ministry of Agriculture, Fisheries and Food, Chief Scientist's Group, 7th Floor, 1A Page Street, London SW1P 4PQ*

C. Hart, *Weetabix Ltd., Weetabix Mills, Burton Latimer, Kettering NN15 5JR*

P. J. Lillford, *Unilever Research, Colworth Laboratory, Sharnbrook, Bedford MK44 1LQ*

A. Penman, *Unilever, Colworth House, Sharnbrook, Bedford MK44 1LQ*

T. H. Slone, *Department of Molecular and Cell Biology, University of California, Berkeley, CA 94720, USA and Department of Molecular and Cell Biology, Ernest Orlando Lawrence Berkeley National Laboratory, Berkeley, CA 94720, USA*

L. Swirsky Gold, *Department of Molecular and Cell Biology, University of California, Berkeley, CA 94720, USA and Department of Molecular and Cell Biology, Ernest Orlando Lawrence Berkeley National Laboratory, Berkeley, CA 94720, USA*

D. M. Taylor, *SEDECON 2000, 147 Oxgangs Road North, Edinburgh EH13 9DX*

J. Wright, *Environmental Resources Management, 8 Cavendish Square, London W1M 0ER*

Assessing the Safety of GM Food Crops

ANDREW CHESSON

1 Introduction

None of the crops currently used as food plants resemble their wild progenitors; all have been modified over the centuries to improve aspects of their quality and productivity. Selection of new traits initially was a pragmatic reaction to random mutational events that gave rise to 'sports', crop plants that differed in some way from the norm. With the development of an understanding of Mendelian inheritance, crosses could be made in a more deliberate and systematic manner and plant breeding and the development of new varieties became a more directed activity. The existing variation within the genetic stock proved inadequate and so irradiation and chemical methods were used to introduce random and often multiple mutations to provide greater diversity from which selections could be made and novel traits introduced. However, all such developments remained within a single species or at least a closely related group of species. Genes were modified or deleted but no new genetic material was introduced. Natural outcrossing was prevented by the highly effective physical and chemical barriers present in plants that restrict cross-fertilization and, in most instances, prevent hybridization.

It was not until the 20th century that species barriers finally were breached, allowing introduction of genetic material from a genetically unrelated plant. The crucial technological breakthrough was not the use of recombinant DNA technology, but the development of tissue culture methods. Tissue culture enabled various single-cell fusion methods to be used which overcame the natural barriers operating at the whole plant level. Such techniques are now considered to fall within the broad description of GM technology and have, retrospectively, been included in the legislation governing the use and introduction of products of GM technology. Modern recombinant DNA methods, the ability to excise a gene from a donor genome and introduce it into a recipient organism in a manner that allows its expression, has now largely made these earlier and far less selective fusion techniques redundant.

Some have argued that recombinant DNA technology applied to crops is

Issues in Environmental Science and Technology No. 15
Food Safety and Food Quality

simply a logical continuation of the manipulations applied since plants were first deliberately cultivated for food use. As such, the novel phenotypes produced are no different in principle from the varieties produced through conventional breeding and should be considered in the same manner. However, once the species barrier was crossed it became possible to introduce genetic material coding for products that have never been part and, with conventional breeding, never could be part of the food supply. For such traits, the security provided by a long history of use by humans is not available and the recognition and knowledge of safe levels of the natural toxicants present in most foods is not automatic. For this reason alone it is necessary to establish regulatory systems able to recognize any hazard introduced by foreign genes and assess the associated risks. Industry concerns that over-stringent requirements for evidence of safety will stifle development of a vital technology have limited validity. It is better to start with rigorous requirements for safety evaluation and then to relax the specifications in the light of experience, rather than having to tighten regulations once evidence of damage to human health has occurred.

The Risk Analysis Framework

Most assessors of risk base their approach on the framework set out in the 1997 FAO/WHO Consultation on Risk Management in which risk analysis is seen to consist of three interlocking components:

- Risk assessment
- Risk management
- Risk communication

The identification and characterization of possible hazards is the starting point for risk assessment which, when coupled with the likelihood of occurrence or exposure, gives a measure of risk. Whenever possible this risk is quantified, although this has proved difficult with many of the hazards suggested for GM foods because of difficulty in establishing measurable outcomes. Risk management then considers how any assessed risk could be reduced, either by removal of the hazard or by a reduction in exposure. Finally, in an ideal and wholly rational world, the process is completed by a cost–benefit analysis in which economic, social and ethical issues are considered and introductions weighed against the cost of inaction. As has become abundantly clear during the GM food debate, since this process requires value judgements, it is essential that those most affected are directly involved in this decision making process.

Recombinant DNA technology is described as a precise tool in which only the intended genes are introduced into the modified host. If this were the only issue then the process of safety assessment would be made easier since only the transgene(s) and their expressed product(s) would need to be considered. Unfortunately, while the nature of the vector and the tDNA may be closely defined, the same is not true of the manner and extent of its incorporation. With the present methods of transformation there is virtually no control over where in a genome the tDNA is introduced, how many copies are integrated, or whether introductions involve the entire gene sequence or just fragments of a gene. Since random insertions could disrupt and silence existing genes or lead to expression of otherwise non-expressed sequences, much has been made of the need to take account of possible 'inadvertent' and therefore unpredictable effects in any safety assessment. While this is a potential hazard, it should be recognized that unintended effects are just as much a problem in conventional breeding as with GM technology.[1] Inadvertent effects have been documented for conventional crops on many occasions without triggering a need for a safety assessment of all new varieties being introduced.

2 GMOs already in the Food Chain

It has been estimated that some 60% of manufactured food items in American supermarkets contain genetically modified ingredients. The equivalent figure for the UK, before food manufacturers and retail outlets began deliberately to source non-GM ingredients, was somewhat lower at 40% of manufactured items. While the presence of GM ingredients in US products might have been expected, the extensive presence on the shelves of European supermarkets came as a considerable surprise to many European consumers. Few recognized the extent to which products of maize (maize starch or maize gluten) and soybean (soya protein, soya grits, soyabean oil, lecithin) were used in the food industry. Because of their importance to food and animal feed production, both were amongst the first crops to be targeted for genetic modification by the plant breeders in the USA. As can be seen in Table 1, the increase in the planting of GM maize and soybean has been rapid, with seed sales in 2000 in the USA indicating that about half of the total soybean and one-third of the total maize crop will be GM varieties. All of the GM maize and soybean met the safety standards of the American Food and Drugs Administration (FDA), who considered them as equivalent to the conventional crop and so did not require separation or identity preservation. While GM maize and soybean have not been grown to any extent in Europe, they were imported as the raw material or products mixed with conventional sources. The UK imports about 1% and Europe about 10% of the USA maize and soybean production. North Americans, and to a lesser extent Europeans, have been consuming in increasing amounts food products containing GM ingredients for several years without any apparent undue effects. The FDA would claim this as a justification of their approach to safety assessment. Some consumer-interest groups, however, argue that this is using humans as guinea pigs and that it is too soon to judge any long-term consequences.

Although maize and soybean represent the bulk of GM material entering the food chain directly or indirectly via animal feed, various other GM food crops have been grown in lesser amounts, principally in the USA and Canada. These include GM tomatoes, squash and potatoes. While fresh GM tomatoes have been sold in the USA (the so-called Flavr Savr tomato), an equivalent product developed in the UK was sold only as the puree and made solely from imported GM tomatoes and labelled as such. Since its introduction in 1996, over two

---

[1] A. J. Conner and J. M. E. Jacobs, *Mutat. Res.*, 1999, **443**, 223.

Table 1 Global area of transgenic crops under cultivation. In 1996 the total area grown was 1.7 million hectares[a]

Commodity	*Area of transgenic crops* (million hectares)		
	1997	*1998*	*1999*
Soybean	5.1	14.5	21.5
Maize	3.2	8.3	11.2
Cotton	1.4	2.5	3.7
Rapeseed (canola)	1.2	2.4	3.4
Potatoes/squash/papaya	0.1	0.1	0.1
Total area	11.0	27.8	39.9

[a]Data from James.[2]

million cans of this heat-treated product were sold before being withdrawn in 1999 by the producers in response to the general flight of UK supermarkets from GM produce.

Rapeseed (canola) is also a likely source of GM material. Although most rapeseed oil is of European origin and from non-GM varieties, some seed is imported from North America for sowing and for use in oil extraction. Most comes from Canada, where use of GM oilseed rape is commonplace (62% of the total crop in 1999). Imports into Europe of non-GM hybrid canola seeds from Canada were discovered in 2000 to contain some GM material (<1%). In the view of some member states, this represented an unapproved release of a GM crop to the environment and several thousand hectares of rape, principally in the UK but also in Sweden and Germany, were ordered to be destroyed. It is unclear whether this mixing of GM with conventional seeds occurred at the seed station or whether there was cross-contamination in the field. However, with the level of use of GM technology in North America, strict segregation will be virtually impossible to maintain and some adventitious contamination of conventional crops is inevitable. Studies originating in the USA have already detected traces of GM material in the majority of consignments of conventional maize tested.[3] If European governments continue a policy of no latitude and to require crops containing trace amounts of GM material to be destroyed, sourcing of seeds from North America will become increasingly impractical.

In addition to ingredients, various additives and processing aids are also derived from GM organisms, most notably some enzymes used in food production. Virtually all hard cheeses are now produced with the aid of a chymosin, an enzyme of bovine origin, cloned into and produced by a bacterium (*Escherichia coli*), a yeast (*Kluyveromyces lactis*) or a filamentous fungus (*Aspergillus niger*). These three sources of nature-identical enzyme replaced the use of bovine rennet and were welcomed by the Vegetarian Society.

Genetic Modifications to Food Plants

It is easy to forget that, even in the USA, the first GM crops were only authorized

[2] C. James, *Global Status of Commercialized Transgenic Crops: 1999*, ISAAA Briefs No 12, ISAAA, Ithaca, NY, 1999.
[3] A. Coghlan, *New Sci.*, 27th May 2000, **166**, 4.

for general release in 1993 and grown commercially in 1995, although such releases were preceded by several years of field trials. Regulatory authorities have had only around 10 years and a handful of practical examples with which to develop a safety assessment strategy. To date, less than 20 structural genes have been introduced singly or in combination into GM plants for which approval is sought for sale or growth in the EU, or whose products can be found in European food. The US experience would extend to perhaps twice this number of genes. Structural genes are also accompanied by regulatory sequences, which direct the expression of the structural gene in the plant. Although most attention has been paid to the safety of the products of structural genes, it has been suggested that some regulatory sequences of viral origin may also pose hazards,[4] although most dispute this.[5]

The structural genes to be found in transgenic plants already authorized for release in at least one country can conveniently be considered in six groups according to their function: insect resistance, herbicide tolerance, virus resistance, male sterility, gene silencing and selection markers.

Insect Resistance. Insect resistance is conferred by genes coding for a truncated form of the protein endotoxins produced by strains of a soil bacterium, *Bacillus thuringiensis* (Bt). Many forms of these toxins exist in nature, affecting different insect species. Maize and tomato crops engineered for insect resistance carry genes of the *cry1(A)* or *cry9(C)* type whose products are specifically toxic to lepidoptera like the European corn borer, while, in potato, incorporation of *cryIII(A)* provides protection against the Colorado beetle. Suspensions of bacteria producing Cry1A toxins have been used as crop protection biopesticide sprays for nearly 40 years, for both glasshouse and field application. Use of this spray is permitted for organic farmers who now are particularly concerned about resistance developing as a result of increasingly wide-scale and continuous use. However, *Bacillus thuringiensis* is a close relative of *Bacillus cereus*, the human enteropathogen, and produces the same human enterotoxins in addition to its insecticidal properties,[6] making its continuing use as a biopesticide open to question.

Herbicide Tolerance. Crops expressing tolerance of one of two herbicides, glufosinate ammonium (Basta) and glyphosate (RoundUp), are the most commonly encountered. Examples of both have sought release in Europe. Elsewhere, plants expressing tolerance of bromoxynil, sulphonylurea and imidazolinone have received market approval. Glyphosate acts by inhibiting an enzyme involved in the synthesis of aromatic amino acids in the plant. Resistance is provided by a bacterial gene coding for the same enzyme which is sufficiently similar to replace the action of the plant enzyme allowing the production of amino acids, but sufficiently different to resist the effect of glyphosate. A maize line made resistant to glyphosate because of a deliberate mutation to the plant enzyme has also been developed and marketed. Glufosinate ammonium is an

---

[4] M. W. Ho, A. Ryan and J. Cummins, *Microb. Ecol. Health Dis.*, 2000, **11**, 194.

[5] J.-B. Morel and M. Tepfer, *Biofutur*, 2000, **20**, 32.

[6] P. H. Damgaaerd, H. D. Larsen, B. M. Hansen, J. Bresciani and K. Jorgensen, *Lett. Appl. Microbiol.*, 1996, **23**, 146.

analogue of L-glutamic acid and a potent inhibitor of glutamine synthase. In most cases, resistance is provided by the introduction of genes (*bar* or *pat*[7]) encoding phosphinothricin acetyl transferases which catalyse the acetylation of glufosinate ammonium, rendering the compound inactive.

Virus Resistance.   It has been known for many years that inoculation of plants with a mild strain of a virus can prevent infection by more virulent strains. Similar effects can be obtained by introducing genes coding for the virus coat protein alone. This was first demonstrated for the tobacco mosaic virus (TMV) in laboratory experiments[8] and subsequently in field trials in 1988 made with tomatoes modified to express TMV coat protein. Virus-resistant varieties of squash, papaya and potatoes expressing the coat proteins of a number of different viruses or other viral proteins (replicase genes) now have market approval in North America. None, however, have sought approval in Europe to date.

Male Sterility.   A variety of mechanisms have been developed to introduce male sterility. This has been dubbed 'terminator technology' because, in some forms under development, genes are included that act as molecular switches able to restore fertility, thus enabling companies to protect their intellectual property rights (see page 16). Considerable concern has been voiced about the impact of this technology on the ability of farmers in poorer regions of the world to save seed for replanting. In practice and to date, the technology has found only benign application in preventing self-fertilization and ensuring uniform hybrid seed production. In these cases, male sterility is introduced by *barnase*, a gene that codes for a ribonuclease and is controlled by a regulatory sequence that allows expression only in pollen. The presence of *barnase* renders the pollen sterile and prevents fertilization.

Gene Silencing.   In addition to introducing genes expressing novel traits, recombinant technology can be used to inhibit or 'silence' the expression of existing, undesirable plant genes. This is always based on the introduction of an extra copy of the target plant gene, either in reverse orientation (antisense) or in the same orientation but in a truncated form (sense). The presence of the additional copy blocks the normal process of gene expression and leads to the suppression of the target enzyme. Probably the best known example is the tomato with increased self-life, in which the tomato enzyme polygalacturonase partly responsible for softening is inhibited. In the case of the Flavr Savr tomato introduced in the USA, the antisense gene was introduced. The risk of patent infringement prevented the UK producer taking the same approach, and so a truncated sense sequence was used for their product. Other applications with, or seeking, market approval are potatoes for starch production in which the granule-bound starch synthase is inhibited, producing starch with different properties of value to industry, and high oleic acid soybean in which a desaturase gene is silenced.

[7]   A. Wehrmann, A. van Vliet, C. Opsomer, J. Botterman and A. Schulz, *Nat. Biotechnol.*, 1996, **14**, 1274.
[8]   P. Powell-Abel, R. S. Nelson, B. De, N. Hoffman, S. G. Rogers, R. T. Fraley and R. N. Beachy, *Science*, 1986, **232**, 738.

Selection Markers. The presence of antibiotic resistance markers in GM constructs has received a great deal of attention because of the concern about the growing resistance of pathogenic bacteria to the antibiotics used in their control. In reality, antibiotic resistance markers in plants do not pose any serious issues of safety. This is partly because the force of public concern already has ensured that alternative methods will be used to select for the introduced traits in GM modified plants in the future (see page 14). In the interim, the two most commonly used selection markers, *nptII* and *aadA*, which code for kanomycin/neomycin and streptomycin resistance, respectively, are antibiotics of very limited importance in human medicine. The remote possibility that either of these resistance genes could transform and be expressed in bacteria found in the digestive tract is of no significance against the background of resistance already present. Use of other antibiotic resistance markers such as ampicillin resistance (*bla*) is less desirable and, however slight, the risk associated with the introduction of amikacin resistance (*nptIII*), a reserve antibiotic of some importance, is unnecessary and unacceptable.

Regulatory Framework for GM Food Crops

Despite some differences of approach, there is considerable agreement over the way in which the various national regulatory authorities approach the question of safety assessment of novel foods, particularly those involving recombinant technology. This harmonization of approach has been aided by discussions in various international fora including the OECD, the WHO and, more recently, through the joint FAO/WHO Codex Alimentarius Commission Task Force on Foods Derived from Biotechnology. There are, however, regional and national differences in the level of risk considered acceptable and this is reflected in the specific provisions of the national legislation. There are also different approaches to food legislation in general, particularly whether the ultimate responsibility for the safety assurance is considered to rest with the national authorities or with the producer.

In the USA, by far the largest producer of GM crops, there is no statutory provision or regulation dealing specifically with food produced by biotechnology. The Food and Drug Administration (FDA), after examining a number of GM products, decided that foods developed through genetic modification are not inherently dangerous and, except in rare cases, should not require extraordinary premarket testing and regulation.[9] The FDA considers that the onus of establishing safety ultimately rests with the petitioner, the consumer being adequately protected by existing regulations which authorize the seizure of adulterated food and enables those responsible for its manufacture and distribution to be prosecuted. The FDA does, however, consult with and provides extensive guidance to industry that describes a standard of care for ensuring the wholesomeness and safety of food developed by rDNA technology. This is in contrast to the European Union where virtually no GM crops are currently being grown, but where specific regulations relating to rDNA technology are in force

---

[9] FDA (Food and Drug Administration), *Statement of Policy: Foods Derived from New Plant Varities*, *Fed. Reg.*, 1992, **57**, 22984.

Table 2 Legislation in force in the European Union controlling the testing, production and release of genetically modified materials

Legislation	Scope of the legislation
Directive 90/219/EEC	Regulates the contained use of genetically modified microorganisms for research and industrial purposes
Directive 90/220/EEC[a] (revised 2000)	Regulates the deliberate release into the environment of genetically modified organisms, including placing on the market of GMOs but not products derived from GMOs. The Directive in its most recent amendment also makes mandatory labelling indicating the presence of GMOs
Regulation EC/258/97	Novel Foods and Novel Food Ingredients Regulation. This sets out the rules for the authorization and labelling of GMO-derived food products
Directive 98/95/EC	Requires that seeds for sale within the EU satisfy the requirements of 90/220/EEC and, if intended for food production, 258/97 before inclusion in the Common Catalogue. Labelling must show that the seeds are of a GM variety
Directive 90/679/EEC	Regulates exposure of workers to biological agents including GMOs
Regulation EC/1139/98	Serves as a model for labelling within the EU and for the requirements for food labelling required by Regulation EC/258/97. In the case of GM foods, labelling provisions are amended by Commission Regulation 49/2000 to provide a 1% adventitious contamination level below which the presence of GM material does not have to be declared
Regulation 50/2000	Deals with the specific labelling requirements for foodstuffs and food ingredients containing additives and flavourings derived from GMOs

[a]Currently no legislation exists dealing with animal feeds although a feed equivalent to the Novel Foods Regulation is in preparation. GM feed issues are dealt with under 90/220/EEC.

requiring a producer to provide evidence of safety (Table 2). The regulations are backed by guidelines that detail the evidence needed. It would be difficult to argue that any producer who had met in full the requirements of these guidelines and whose product had been authorized had acted with anything less than due diligence. Consequently, the Commission is, in effect, assuming responsibility for safety assurance within the EU.

Assessment of GM Food Crops for Human Safety

Substantial equivalence. The much maligned term 'substantial equivalence' is widely accepted as the starting point for the human safety assessment of GM products. An equivalent concept of 'familiarity' is used in the assessment of environmental safety. The concept of substantial equivalence derived from the

FDA view that the first GM crops gave rise to new varieties that were not, in principle, different from other varieties produced by 'conventional breeding'. All were considered to be substantially equivalent (but not identical). This became embodied in the OECD statement 'that existing organisms used as food or as a source of food, can be used as the basis for comparison when assessing the safety of consumption of a food or food component that has been modified or is new'.[10] However, the FDA recognized that, as GM technology developed, crops that differed radically from their conventional counterparts could be produced. From this developed three scenarios intended to provide a flexible framework for safety assessment and three different 'gateways' to assessment protocols:

1. When substantial equivalence can be established for an organism or food product, it is considered to be as safe as its conventional counterpart and no further evaluation is needed.
2. When substantial equivalence can be established, apart from certain defined differences, further safety assessment should focus on these differences.
3. When no conventional counterpart exists or where substantial equivalence cannot be established, the design of any test programme should be on a case-by-case basis taking into account the characteristics of the food or food component. A failure to establish substantial equivalence does not automatically mean that a product is unsafe or even that extensive safety testing will be required.

Substantial equivalence is not and was never intended as evidence of safety *per se*. It is the product of a number of measures, usually based on a chemical analysis, each of which provides specific evidence of safety. However, neither 'substantial' nor 'equivalence' have dimensions and the number and nature of tests required to establish substantial equivalence is not fixed. The tests applied are determined partly by the known characteristics of the conventional counterpart, particularly the presence of known toxicants, and partly by the nature and manner of the genetic modification and the degree of presumed hazard.

Unfortunately, the concept of substantial equivalence has been misapplied to the question of *inadvertent effects*. These are any unexpected consequences of the insertion of foreign genetic material which may arise either because the expressed product distorts the flux of metabolites in a manner not predicted or that the random insertion of the tDNA alters the expression of an existing gene or group of genes. Chemical analysis may reveal a significant and unexpected difference between conventional and GM plants, pointing to an inadvertent effect. However, the absence of differences in a necessarily limited number of analyses used to establish substantial equivalence does not prove the absence of an inadvertent effect. If inadvertent effects are considered a hazard requiring greater attention, then other methods of detection will need to be developed (see page 22).

Safety Assessment of tDNA. All of the 40 or so GM crop varieties examined and authorized for release in at least one country to date have fallen under the second

---

[10] OECD, *Safety Evaluation of Foods Derived from Modern Biotechnology: Concepts and Principles*, OECD, Paris, 1993.

category described above. In each case an appropriate conventional plant was available for comparison and the expected agronomic and nutritional characteristics for the species well established. This enabled substantial equivalence to be reasonably concluded and the bulk of the safety assessment then focused on the nature and expression of the inserted DNA. As a starting point, all regulatory authorities require the origin and nature of all genetic elements that have been introduced, including all regulatory sequences and any remaining parts of the vector sequence, to be precisely defined. All open reading frames have to be identified and host sequences flanking all inserted DNA must be subject to nucleotide sequence analysis to determine if the inserted DNA has the potential to generate novel fusion proteins. Once precise knowledge of the insertional event is established, the possibility of horizontal gene transfer and its consequences can be considered. For GM plants this includes the important issue of the extent of transfer of genetic material to other varieties of the same crop plant or to other compatible species. Once ingested, the possibility of gene transfer to the microflora of the gut may be of concern, particularly where antibiotic resistance markers are present in the final construct. Such concerns may also extend to the soil microflora.

Once the identification of any hazards associated directly with genetic material has been completed, knowledge of which new proteins are expressed in the transformed plant is needed. This includes both their concentration and localization, as not all plant parts are consumed. Each new protein then needs to be assessed for potential toxic effects and for any allergic or delayed sensitivity reactions arising from exposure. Toxicity testing, which requires the test article to be fed at concentrations substantially greater than would be delivered by the whole food or feed, is based on internationally approved acute and chronic studies in laboratory animals or fast-growing domesticated species such as chickens. Finally, the consequences for the plant metabolism may have to be assessed, depending on the nature of the genetic modification made. Herbicide resistance, for example, may result in novel metabolites of the herbicide following its application. This in turn may require a re-assessment of any herbicide residues in the tissues of humans and animals consuming the treated crop.

Food Labelling

Labelling Requirements. Food labelling is seen in most countries as a valuable source of information for consumers, enabling them to make a more informed choice when selecting products. However, until recently there was no requirement for the labelling to indicate a GM origin in the United States and this remains the case in many other countries. Even in the US, following a statement from the White House in May 2000, the FDA was charged only with developing guidelines for the *voluntary* labelling of food products as containing or not containing bioengineered products. This is in marked contrast to most European countries, particularly the EU member states, where positive labelling for GM content is mandatory (Table 2).

Any positive labelling requirement has to take account of what is described in the EU as 'adventitious contamination', the almost inevitable accidental mixing of commodities that can occur at all stages in a distribution and manufacturing process. It is normal to set a *de minimis* value, below which a declaration is not required. This value is usual arrived at pragmatically, taking account of the level of contamination likely, the degree to which contamination represents a health hazard and the availability of robust methods of detection available to enforce the legislation. In the EU the requirement is that no more than 1% of the product should be of GM origin. Many consumer organizations would also like to provision for negative labelling, *i.e.* that a product can be declared GM-free. This would also require setting of a second *de minimis* level, possibly around 0.1% GM content, below which the product could be said to be GM-free. However, although PCR and antibody-based methods capable of this level of detection exist, they are not universally effective, particularly in food products that have been extensively processed.[11] Consequently, few countries have provision for negative labelling, although the European Commission is preparing a proposal for a Regulation on the labelling of GM-free foodstuffs.

Negative Lists. Opinions differ on whether products of GM crops intended for food use which are essentially free of DNA and protein should be treated in the same manner as those derived from their conventional counterpart and whether labelling should indicate their GM origin. Such products currently include the refined and cold-pressed oils from soybean, maize, rape and cotton seed and starch extracted from GM potatoes and its hydrolysis products. For most countries the concept of substantial equivalence applies and they are treated as any other oil or starch product and are not subject to safety assessment or labelling requirements. In Europe, the European Commission proposed to establish a negative list of food products essentially free of DNA and protein that would be exempted from labelling provisions. However, on examination, the presence of DNA was found in virtually all cold-pressed oils and in some refined oils, although the amount in refined oils was small and could only be detected by nested PCR amplification.[12,13] Since, the term 'highly refined oil' could not be unambiguously defined or any one process identified which removed all DNA/protein, this attempt was abandoned. As a result, all such products currently are captured by existing EU legislation and require labelling.

Post-market Surveillance/Monitoring. A number of European countries are strong advocates for some form of post-market surveillance or monitoring of GM products released for sale. This is made possible by product labelling and the newly amended EU Directive 90/220/EEC uniquely makes provision for this. They are, however, two serious problems associated with this approach, the first relating to public confidence and the second to its practicality. Any provision for post-market surveillance casts doubts on the safety assessment already completed and implies a degree of uncertainty that warrants an extensive and expensive surveillance exercise. This, it has been argued, will inevitably undermine whatever public confidence remains in the science-based assessment process adopted by

[11] H. A. Kuiper, *Food Control*, 1999, **10**, 339.
[12] M. Hellebrand, M. Nagy and J.-T. Mörsel, J-T. *Z. Lebensm. Unters. Forsch. A.*, 1998, **206**, 237.
[13] TNO, *Soy Oil*, Report 389/352, TNO, Voeding, The Netherlands, 1999.

A. Chesson

the various national authorities. The second difficulty is that many hundred million North Americans have consumed GM foods for several years and no adverse effects associated with GM food consumption have been identified. If there are any untoward effects of GM foods, they must be chronic in nature and of very low incidence or very long gestation. A key requirement for establishing a system for surveillance is that a hypothetical outcome exists capable of being detected. Even when a testable hypothesis exists, evidence can be difficult to obtain. The epidemiological evidence linking smoking with lung cancer took many years to establish despite a clear measure of exposure and a defined and well-documented outcome. In any diet-based prospective study, good consumption data for individuals is difficult to obtain, as it may be difficult to assess the GM content of the diet and, since present GM products are thought safe, by definition no logical outcome of concern can be defined.

Animal Feedstuffs

Most animal feeds, particularly manufactured feedstuffs, make use of the same crops (or by-products of the same crops) used for human food. As such, GM crops, whether entering the human food chain directly or via animal feed, are assessed in much the same way and often conjointly. If animal use is intended, such safety assessments additionally take account of any risk to the target species (the species of livestock) and the indirect risk to the consumer of animal products (meat, milk and eggs) of factors such as possible harmful residues.

There are some aspects specific to feed use that need to be considered when assessing GM material for animals. Most safety assessments for human foods are made with that part of the plant consumed in unprocessed form and the results assumed to apply to any products derived from this source. Animals, however, are often uniquely fed by-products of processing in which the DNA/protein, including the tDNA and the products of its expression, are substantially concentrated. Thus, the assessment of safety may be better made with the by-product as fed, rather than with the whole crop or part of the crop used for human food. Conversely, the methods used for feed processing may ensure the (partial) disruption of DNA and protein.[14,15] Animals also consume parts of the plant not fed to humans, such as whole crop maize fed ensiled to dairy cattle. With the commonly employed 35S promoter from the cauliflower mosaic virus, expression of herbicide tolerance or insect resistance in the green tissues is often 10- to 100-fold greater than that found in seeds or pollen. This level of expression has to be taken into account when developing toxicity assays. Other feed ingredients may never enter the human diet directly. Ruminant animals, because of their large resident microflora, are able to degrade many plant materials toxic to non-ruminants, including humans. Thus GM cotton enters the food chain not primarily as cottonseed oil, but because most of the cottonseed meal left after extraction is fed to dairy cows capable of degrading gossypol, the major toxicant present.

[14] D. E. Beever and C. E. Kemp, *Nutr. Abstr. Rev. Ser. B*, 2000, **70**, 175.
[15] N. Smith, E. R. Deaville, W. S. Hawes and G. C. Whitlam, in *Recent Advances in Animal Nutrtion 2000*, ed. P. C. Garnsworth and J. Wiseman, Nottingham University Press, Nottingham, 2000, p. 59.

12

Unlike foods specifically for human use, genetically modified feedstuffs can be fed directly to the target species and the outcome measured, whether in terms of growth, health and welfare, or to determine absorption and tissue distribution of particular metabolites. Although the limits to the amount of any one component that can be incorporated into an animal's rations usually prevent chronic toxicity studies being made with the target species, the wholesomeness of the product can and should be directly demonstrated. The production behaviour of the animal on the GM diet when compared to a conventional diet also provides weight in establishing substantial equivalence.

3 Assessing Environmental Safety

This review has as its focus the safety of GM foods for the consumer. However, it is recognized that safety for the wider environment is a major component of the risk assessment applied to GM crop plants and the subject of much debate by professional environmentalists, pressure groups and by the public generally. Serious concerns have been expressed about the ability of a GM crop to cross-pollinate adjacent areas of conventional crops of the same species, and the extent to which out-crossing will occur with other related species. Some crops already selected for modification, notably the brassicas, are widely distributed and hybridization with related species is known to occur.[16] Many of these concerns have focused on the uncontrolled spread of herbicide resistance. Distances between crops necessary to maintain the genetic purity of varieties are found in the legislation governing seed production in most countries and provide a guide to what is known about pollen dispersal. These distances vary depending on the degree to which the crop is self-fertilized and whether cross-fertilization is wind-dependent. However, the level of purity defined for seed varieties (99–99.9% for *Brassica* spp. in the UK) may not be adequate to meet the needs of particular interest groups.

The development of resistance to the Bt group of toxins is another issue of concern[17] since application of *B. thuringiensis* cells, or extracts of the organism as a biocide, is one of a few insecticidal treatments organic farmers are allowed. It has yet to be shown whether the provision of refuge areas, now a statutory requirement, are adequate to prevent build-up of resistance. Other concerns relate to effects of plants in which Bt toxin is expressed on non-target species. These were triggered by various laboratory-based studies on the survival of the Monarch butterfly[18] and a number of other lepidoptera, including beneficial predators[19,20] apparently susceptible to Bt toxin-containing pollen. However, subsequent work has shown most of these concerns to be unfounded, or to

[16] J. A. Scheffler and P. J. Dale, *Transgenic Res.*, 1994, **3**, 263.

[17] Union of Concerned Scientists, *Now or Never: Serious New Plans to Save a Natural Pest Control*, ed. M. Mellon and J. Rissler, UCS Publications, Cambridge, Mass., 1998.

[18] J. E. Losey, L. S. Raynor and M. E. Carter, *Nature*, 1999, **399**, 214.

[19] A. Hilbeck, W. J. Moar, M. Pusztai-Carey, A. Filippini and F. Bigler, *Environ. Entomol.*, 1998, **27**, 1255.

[20] A. Hilbeck, W. J. Moar, M. Pusztai-Carey, M., A. Filippini and F. Bigler, *Entomol. Exp. Appl.*, 1999, **91**, 305.

have a far lesser impact than the alternative use of insecticides for control purposes.[21,22]

Perhaps less well recognized is the fact that concerns seen as the preserve of the environmentalist may also have a bearing on human health and safety. The cross-fertilization between a GM crop approved for human food use and a conventional variety of the same species may impact on the livelihood of an organic farmer but is unlikely to have any consequence for human health. Cross-pollination involving a GM crop approved for non-food use, however, may have a more profound outcome. It will become increasing important to develop risk management strategies to avoid transfers from potentially dangerous constructs. Bt toxin expressed in maize can significantly reduce insect damage and so lessen the conditions that encourage fungal infection. In consequence, mycotoxin production, particularly those produced by *Fusarium* spp. (fumonisin, vomitoxin, monoliformin), can be substantially reduced to the benefit of both humans and livestock.[23,24] Expression of the Bt toxin, which is toxic only to insects, also avoids any dangers associated with the use of *B. thuringiensis* itself for biological control because of potential risks of human enterotoxin production.[6]

4   Improved Risk Management through Developments in Recombinant DNA Technology

Recombinant DNA technology is rapidly evolving and new methodologies are being constantly introduced. It is to be expected that many of these will help to offset concerns that surround the present generation of GM products, either by finding more acceptable alternatives or by providing management strategies to reduce an identified risk.

Selection Markers

Questions raised about the use of genes expressing antibiotic resistance as a means of selecting transformed plants have led to alternative strategies being developed. Where antibiotic-resistance genes are still used for the screening of transformants, techniques are available to remove these from the selected transformed lines. These include site-specific deletions, transposon-mediated relocations and the co-transformation of marker genes on separate vectors. However, all of these processes are far from straightforward and may not be suitable in all cases.

Selection systems serve two functions, firstly reporting the presence of a successful transformation and, secondly, providing the means of rapidly isolating successful events from thousands of transformants. Antibiotic resistance markers combine both functions in a single gene by allowing only transformed cells to survive in the presence of an otherwise lethal concentration of the antibiotic.

[21] D.S. Pimentel and P.H. Raven, *Proc. Natl. Acad. Sci. USA*, 2000, **97**, 8198.
[22] NRC (National Research Council), *Genetically Modified Pest-Protected Plants: Science and Regulation*, National Academy Press, Washington, 2000.
[23] G.P. Munkvold, R. Hellmich and W.B. Showers, *Phytopathology*, 1997, **87**, 1071.
[24] G.P. Munkvold, R. Hellmich and L.G. Rice, *Plant Dis.*, 1999, **83**, 130.

Alternative marker systems exist and, as these are improved, it is only a matter of time before the use of antibiotic marker genes will be phased out entirely. This process may be accelerated by future legislative requirements. Herbicide resistance provides one mechanism directly analogous to antibiotic resistance. Obviously suited to those plants intended to demonstrate herbicide resistance in the field, herbicide resistance can be used for other events provided that the levels of resistance introduced is adequate only for selection purposes and would not be significant in the field. As an alternative, plants naturally lacking a capacity to utilize xylose or mannose as a sole carbon source can be selected by the introduction of genes coding for mannose-6-phosphate isomerase or xylose isomerase, providing a system in which only transformed cells can develop. Selection and reporting do not have to reside with a single gene; dual systems also exist, such as the use of cytokinin glucuronides as the selective agent and β-glucuronidase as the selectable marker.

Control over the Location of Gene Insertion

Possibly the greatest uncertainty in terms of public health is the lack of control over where in a recipient genome the introduced DNA is inserted. This is particularly the case when biolistic methods are used to transform plant cells. Following the successful construction of a yeast artificial chromosome (YAC) and the demonstration that it was maintained through successive generations of yeast,[25] similar proposals have been developed for a plant artificial chromosome (PAC). If successful, PACs promise to be powerful candidates for future vectors in plant transformations.[26,27] PACs would permit many genes to be introduced simultaneously into plants in order to engineer a complex feature such as a novel pathway or the amino acid content of seeds. In addition, they would allow the introduced genetic material to be precisely defined and to be introduced and stably maintained within the plant without any disruption to existing genetic elements. However, an important factor in the successful development of the YAC was size. It was possible to assemble a molecule large enough to contain all of the elements of a functioning yeast chromosome, but small enough such that only minor modifications to conventional recombinant DNA techniques were required. This will not be the case with a PAC, which will have to be substantially larger.[28]

A serious alternative to existing or proposed transformations of the nuclear genome is the modification of plastids.[29] Plastids are one of the three gene-containing compartments in the plant cell, the others being the nucleus and the mitochondria. They occur in several forms including chloroplasts, carotenoid-containing chromoplasts, starch-containing amyloplasts and oil-containing elaioplasts. Unlike the nuclear genome, the chloroplast chromosome typically contains only 120–150 genes, most organized into operons under a common

[25] A. W. Murray, N. P. Schultes and J. W. Szostak, *Cell*, 1986, **45**, 529.

[26] T. Yamada, *Seibutsu Kogaku Kaishi*, 1998, **76**, 205.

[27] W. R. A. Brown, P. J. Mee and M. H. Shen, *Trends Biotechnol.*, 2000, **18**, 218.

[28] D. Shibata and Y.-G. Lui, *Trends Pharm. Sci.*, 2000, **5**, 354.

[29] L. Bogorad, *Trends Biotechnol.*, 2000, **18**, 257.

regulatory control. Biolistic methods or direct microinjection can be used to introduce the tDNA,[30] which then is incorporated into the chloroplast chromosome by homologous recombination.[31] Thus tDNA can be directed to a specific site, avoiding any inadvertent mutational events. Expression levels may also be substantially higher since the plastids contain multiple copies of the chromosomes and the plant multiple plastids per cell.[32] Plastid engineering may not be appropriate for all applications. It is likely to be the method of choice where multiple genes need to be introduced under common regulation, where the target metabolic activity is centred on a plastid, or where excess product accumulation may be damaging to the cell.

Control of Horizontal Gene Flow between Plants

Although the dangers of transgenic crops interbreeding with nearby weeds and increasing their competitiveness have been overstated, nonetheless there are real concerns about transgenic crops cross-pollinating with conventional counterparts. In addition, GM crops may themselves become 'volunteer' weeds in following crops if both share the same patterns of herbicide resistance, a situation that has already arisen with GM cotton and maize. There is obvious value in developing strategies that fully or selectively prevent horizontal gene transfer which can be applied in circumstances where assessments provide evidence of risk. These can be fail-safe (absolute) in nature or act by transgenetic mitigation to selectively disadvantage weeds or other crops receiving the transgenes.

Fail-safe mechanisms for inducing male sterility exist and have been approved in a number of crop plants where they have been used to assist the production of uniform hybrid seeds by preventing self-pollination. The most common form involves the expression of a specific ribonuclease with expression directed to the tapetum cell layer of the anther, the site of pollen formation.[33] A fertility restorer also exists for this system which inhibits the action of the ribonuclease, allowing normal pollen formation to occur. Attempts to link these and other systems of fertility disruption and restoration together have been pejoratively labelled terminator technology. News of these developments became a rallying cry for various pressure groups, in particular some international aid agencies, concerned about a theoretical loss of a farmer's freedom in developing countries to use farm-saved seed without paying for the means of triggering the fertility restoration gene. However unlikely this scenario, sterility is seen as a potentially oppressive measure rather than what is likely to prove to be an essential element in the health and environmental risk management of GM crops.

Even the chemical-induced switching mechanisms being developed have far greater implications than simply fertility control because they allow more temporal control over gene expression. Insect-resistance genes can be switched on and off with chemical triggers, a mechanism that could be used to obviate

[30] M. Knoblauch, J. M. Hibberd, J. C. Gray and A. J. R. van Bel, *Nat. Biotechnol.*, 1999, **17**, 906.

[31] A. D. Blowers, L. Bogorad, L., K. B. Shark and J. C. Sanford, *Plant Cell*, 1989, **1**, 123.

[32] K. E. McBride, Z. Svab, D. J. Schaaf, P. J. Hogan, D. M. Stalker and P. Maliga, *Biotechnology*, 1995, **13**, 362.

[33] C. Mariani, M. De Beuckler, J. Truettner, J. Leemans and R. B. Goldberg, *Nature*, 1990, **347**, 734.

some of the concerns about the more rapid development of resistance when a gene product is continuously expressed in a plant. However, a better and more immediate approach is to deploy promoters which concentrate expression in those parts of the plant actually attacked by insects, such as a phloem-specific promoter for genes whose products are active against phloem-sucking insects such as aphids.[34] These is also scope for linking expression to wounding, such that the pesticide is produced only when the plant is attacked.[35]

A variety of other actual and theoretical fail-safe systems exist including apomixis, a system being developed in crops to provide hybrid vigour without the need for cross-pollination.[36] In apomixic plants the seed is of vegetative origin and does not derive from pollination. Apomixic varieties either lack pollen or produce non-compatible pollen, preventing gene transfer. The use of plastids in recombinant technology also provides a natural barrier to gene flow since plastid genomes are usually maternally inherited and cannot move into other species. In plants with multiple genomes derived from different sources, more selective measure are possible. Often only one of the genomes is compatible with the wild host such as the D genome of wheat that is compatible with *Aegilops cylindrica*, a problem weed of the USA.[37] The natural integration of a transgene from the A or B genome is much less likely than from the D genome following an interspecies cross and would require a rare homologous recombination.

Transgenetic mitigation, as a control measure, is based on three premises. First, that a tandem construct behaves as a single gene; second, that there are traits neutral to the crop that can disadvantage weeds or other wild species; and third, that because weeds compete strongly amongst themselves and with other species, any mildly harmful trait would be eliminated from the population.[38] Thus, if the desired gene is flanked by a mitigating gene in a tandem construct, effects in the crop would be beneficial but in any other species the effects of the deleterious gene would ensure that the tandem construct was selectively disadvantaged and lost. Various traits to be used in tandem have been suggested, including the abolition of secondary dormancy and dwarfing.

5   Future GM Crops: from Fantasy to Field Trials

Most of the 40 or so transgenic crops with regulatory approval for release have been modified to improve their agronomic properties, particularly by the introduction of herbicide tolerance to allow better weed control and/or some form of pest resistance. Both of these types of traits contribute to obtaining high yields and, coupled with the need to reduce post-harvest losses, are crucial factors in maintaining the security of the world's food supply. Despite the fact that consumers are said to be unable to see a benefit accruing from such developments,

[34] Y. Shi, Y., M. B. Wang, K. S. Powell, E. VanDamme, V. A. Hilder, A. M. R. Gatehouse, D. Boulter and J. A. Gatehouse, *J. Exp. Bot.*, 1994, **45**, 623.
[35] P. A. Lazzeri, in *Genetic Engineering of Crop Plants for Resistance to Pests and Diseases*, ed W. S. Pierpoint and P. R. Shewry, British Crop Protection Council, Farnham, 1996, p. 8.
[36] A. M. Koltunow, R. A. Bicknell and A. M. Chaudhury, *Plant Physiol.*, 1995, **108**, 1345.
[37] R. S. Zemetra, J. Hansen and C. A. Mallory-Smith, *Weed Sci.*, 1998, **46**, 313.
[38] J. Gressel, *Trends Biotechnol.*, 1999, **17**, 361.

their underlying importance is such that developments of such traits will continue to dominate GM crop introductions. However, the sequencing of plant genomes, such as that of rice,[39] and ascribing functionality are beginning to identify the genetic basis for many other traits and, in so doing, providing novel targets for manipulation. Transformations already achieved affect the nutritional, organoleptic and storage qualities of produce; others address public health concerns, the industrial potential of crop-derived products or growth in marginal areas. A few of the many developments in the pipeline are described in this section and, in the following section, the implications for safety assurance of a substantially increased range and number of traits seeking regulatory approval are considered.

Pest Resistance

Recombinant DNA technology has dramatically increased the range of options for engineering resistance to specific insect pests.[40] Although present commercial releases contain only genes derived from *B. thuringiensis* and coding for one of the family of insecticidal δ-endotoxins,[41] some 40 other resistance genes have been introduced into plants. Like the Bt endotoxins, which are lectins binding to carbohydrate structures in the insect gut, most target some aspect of insect gut function. These include various amylase and proteinase inhibitors of plant and animal origin, of which the most potent appears to be the cowpea trypsin inhibitor active against a broad spectrum of lepidopteran and coleopteran species.[42] A number of plant lectins with insecticidal properties have also received attention, although concerns about their potential mitogenicity and toxicity to humans has restricted the range considered to those known to be less toxic to mammals. Of these, the snowdrop lectin (*Galanthus nivalis* agglutinin) has received greatest attention and has been successfully expressed in at least nine different crops, including rice.[34]

A variety of other genes have been identified which target aspects of insect metabolism and development other than gut function. Introduction of a fungal cholesterol oxidase, capable of disrupting membrane integrity, proved highly toxic to larvae of the boll weevil and to the tobacco budworm, both pests of cotton.[43] Chitinases engineered into plants primarily to provide a defence against fungal pathogens also have proved weakly effective against insect pests. Better results have been obtained with peroxidases, although the mechanism of protection is unclear. Another form of protection was provided by expression in tobacco of the coat protein and RNA 1 from the *Helicoverpa armigera* stunt virus, an insect RNA virus, which caused stunting of *H. armigera* larvae.[44] However, it has yet to be demonstrated that the same degree of protection occurs in cotton, the natural host species.

[39] T. Saski and B. Burr, *Curr. Opin. Plant Biol.*, 2000, **3**, 138.

[40] T. H. Schuler, G. M. Poppy, B. R. Kerry and I. Denholm, *Trends Biotechnol.*, 1998, **16**, 168.

[41] M. Peferoen, *Trends Biotechnol.*, 1997, **15**, 173.

[42] A. M. R. Gatehouse and V. A. Hilder, in *Molecular Biology in Crop Protection*, ed. G. Marshall and D. Walton, Chapman and Hall, London, 1994, p. 177.

[43] H.-J. Cho, K. P. Choi, M. Yamashita, H. Morikawa and Y. Murooka, *Appl. Microbiol. Biotechnol.*, 1995, **44**, 133.

[44] T. N. Hanzlik and K. H. J. Gordon, *Adv. Virus Res.*, 1997, **48**, 101.

Quality Traits

Introduction of quality traits has not been high on the agenda of most plant breeding companies since improved agronomic performance was seen both as important and to offer the fastest return on the early investments made in GM technology. There have been notable exceptions, of which the GM tomato engineered for longer shelf-life is the best known example. Following the success of this product, equivalent modifications to the expression of cell wall degrading enzymes involved in the softening of fruits during ripening have been made in many crop products with similar storage problems. Foods with increased physiological functionality, which can deliver some health benefits to consumers, have been widely discussed but relatively little researched. Exceptions include research on hypoallergenic rice, in which genes encoding seed proteins identified as allergenic are down-regulated,[45] and the so-called 'yellow rice' expressing β-carotene, a vitamin A precursor, in the endosperm.[46] The latter example is one of the few occasions when an entire biosynthetic pathway involving four specific enzyme steps has been successfully introduced into a crop plant and the final product expressed, albeit at very low concentration.[47] Yellow rice is designed to contribute to reducing blindness in those regions where rice is a staple and where vitamin A intake from vegetables is low. It is one of a group of rice constructs which also address the incidence of anaemia in women of child-bearing age in these regions by increasing the iron content of rice endosperm[48] and reducing its chelation by phytic acid. Attention has been drawn to the possible health benefits of lycopene which, like carotene, also derives from phytoene. Transgenic tomatoes in which lycopene concentrations are substantially increased have been produced, although apparently this gain is made at the expense of the β-carotene content for which lycopene can act as precursor.

Quality has been more of an immediate issue in crops intended for feed use. Attempts have been made to modify the nutritional value of maize used for silage by reducing or modifying lignin biogenesis[49] and of forage grasses by increasing the soluble carbohydrate content. Protein quality has also been an issue of importance both in green tissues[50] and in seeds.[51] It was attempts to increase the methionine content of soybean, using a gene coding for a methionine-rich protein from the Brazil nut, which first drew attention to the possibility of transferring allergenicity.[52,53]

45 R. Nakamura and T. Matsuda, *Biosci. Biotechnol. Biochem.*, 1996, **60**, 1215.
46 X. D. Ye, S. Al Babili, A. Kloti, J. Zhang, P. Lucca, P. Beyer and I. Potrykus, *Science*, 2000, **287**, 303.
47 P. K. Burkhardt, P. Beyer, J. Wunn, A. Kloti, G.A. Armstrong, M. von Schledz, J. Lintig and I. Potrykus, *Plant J.*, 1997, **11**, 1071.
48 F. Goto, T. Yoshihara, N. Shigemoto, S. Toki and F. Takiwa, *Nat. Biotechnol.*, 1999, **17**, 282.
49 A. M. Boudet, *Plant Physiol. Biochem.*, 2000, **38**, 81.
50 M. R. I. Khan, A. Ceriotti, L. Tabe, A. Aryan, W. McNabb, A. Moore, S. Craig, D. Spencer and T. J. V. Higgins, *Transgenic Res.*, 1966, **5**, 179.
51 L. Molvig, L. M. Tabe, B. O. Eggum, A. E. Moore, S. Craig, D. Spencer and T. J. V. Higgins, *Proc. Natl. Acad. Sci. USA*, 1997, **94**, 8393.
52 O. L. Frick, *ACS Symp. Ser.*, 1995, **605**, 100.
53 J. Nordlee, S. L. Taylor, J. A. Townsend, L. A. Thomas and R. K. Bush, *New Engl. J. Med.*, 1996, **334**, 688.

Industrial Uses

Plants have always be a source of natural, high-value chemicals, predominantly for pharmaceutical or cosmetic use. Recombinant technology has greatly expanded the options for the production of high-value products, particularly peptide- and protein-based therapeutics. Numerous transgenic constructs exist which are able to express a variety of products of potential value in human medicine that can be produced with minimum risk of contamination with human viruses or other disease promoting organisms. These range from human somatotropin in tobacco chloroplasts engineered by plastid transformation[54] to proteins in plant foods able to act as edible vaccines[55] and shown to elicit an immune response when consumed by humans.[56] At present, expression levels generally are inadequate for commercial exploitation and few, if any, have progressed to field and clinical trials. Intermediate value products such as bulk enzymes also share the same problems of poor expression and have yet to challenge existing fermentation processes. However, considerable investment is being made in this area because of the low costs and the ease with which production can be scaled to demand,[57] and there is every reason to suppose that such plant-derived products will enter the market.

At the other end of the scale, low-cost bulk feedstock can also be produced and could replace, in part, feedstock derived from non-renewable sources. Bulk chemical production has focused on the use of oilseed rape in which the oil produced has been modified by changing the expression of key enzymes. One of the first GM plants to be approved for release in the USA (1994) was an oilseed rape modified to produce high concentrations of lauric acid for use in the detergent industry. However, for a variety of reasons, no commercial plantings have been made to date. Since then, other transgenic rape varieties designed for industrial purposes have reached the stage of field trials.[58] Many involve suppression of desaturase genes, resulting in an accumulation of saturated fatty acids such as stearate for food use, but other transformations have led to the production of more toxic seed oils for polymer production or lubricants.

The only economically significant route for the disposal of the seed meal remaining after the extraction of oil from oilseed plants is as animal feed and, at present, all batches of oilseed meals are treated as nutritionally equivalent. It would be a cause of considerable concern if, for example, the lauric acid content of meat or milk were to rise because animals were fed the meal from a high lauric acid rape variety. Lauric acid is known to be a potent stimulator of blood cholesterol in humans. Similarly, erucic acid, a well-known toxicant bred out of conventional rape varieties, has also been over-expressed, reaching concentrations at least 10-fold higher than in any previous rape seeds.[58] Feed producers will need, in the future, to accurately source their raw ingredients to ensure that residues of highly potent biological agents or toxic chemicals are absent.

54 J. M. Staub, B. Garcia, J. Graves, P. T. J. Hajdukiewicz, P. Hunter, N. Nehra, V. Paradkar, M. Schlittler, J. A. Carroll, L. Spatola, D. Ward, G. N. Ye and D. A. Russell, *Nat. Biotechnol.*, 2000, **18**, 333.
55 A. M. Walmsley and C. J. Arntzen, *Curr. Opin. Biotechnol.*, 2000, **11**, 126.
56 C. Tackett, H. Mason, G. Losonsky, M. Esrtes, M. Levine and C. Arntzen, *J. Infect. Dis.*, 2000, **182**, 302.
57 D. Mison and J. Curling, *Biopharm*, 2000, **13**, 48.
58 J. M. Murphy, *Trends Biotechnol.*, 1996, **14**, 206.

Labelling for GM animal feeds probably will not greatly help since it will indicate only that the feed is GM, not the nature of the modification. In the absence of any risk management strategy, traditional routes for the disposal of some crop by-products may have to be reconsidered.

6 Safety Assessment of Future GM Crops

All existing safety assessment schemes make provision for the identification of transgenes and their expression products and for the consideration of their safety in relation to intended use. Each different gene is therefore capable of triggering a series of specific questions relating to the known biological properties of the gene and its product. In this respect, each assessment of a GM crop is unique and all crops are assessed on a case-by-case basis. In theory at least, the introduction of any novel trait should be accommodated by the existing safety assessment processes. However, there are a number of issues which are generic, some of which are not yet fully resolved and others of which are likely to be increasingly challenged by the novelty and complexity of new introductions.

Allergenicity

The capacity to produce an allergic response ranging from a mild sensation of discomfort, through rhinitis to full-blown anaphylaxis, is a common property of existing foods. Approximately 1% of the adult population and 6–7% of children are allergic to one or more foodstuffs that include fish, shellfish, various fruits and nuts, milk, eggs and soybean. Although several hundred allergens have been identified, they have insufficient in common to allow a definitive prediction of allergenicity in a novel protein. At present, no adequate and validated animal models exists which would allow a GM food to be tested for allergenicity.[59] Most existing safety assessment schemes follow the ILSI-IFBC decision tree.[60] This distinguishes between genes obtained from known allergenic sources and those from sources with no known history of allegenicity. Ironically, knowing that a gene originates from an allergenic source provides the greatest security. Very few of the 40 000–50 000 genes that make up an average plant code for allergens and the vast majority of proteins are non-allergenic. Since sensitive patients exist from whom sera can be obtained for immunological assays and who are available for food challenge trials, it is possible to distinguish allergens from non-allergens with a high degree of assurance. The tests allowing proteins from non-allergenic sources to be declared non-allergenic are far less definitive and precise. These are based on the absence of sequence homology with known allergens, evidence of a rapid rate of breakdown in the digestive tract and a low concentration in the finished food or food product. A history of non-allergenicity has relevance only when the gene in question comes from a plant consumed as a food or where there is evidence of some other form of human exposure. Increasingly, genes are being obtained from non-food (and non-plant) sources which may contain allergens

[59] G. F. Houben, L. M. J. Knippels and A. H. Penninks, *Environ. Toxicol. Pharmacol.*, 1997, **4**, 127.

[60] D. D. Metcalfe, J. D. Astwood, R. Townsend, H. A. Sampson, S. L. Taylor and R. L. Fuchs, *Crit. Rev. Food Sci. Nutr.*, 1996, **36**, S165.

with properties other than those expected. In the short term, allergenicity testing could be improved by introducing immunoassays using the pooled sera of patients exhibiting a wide range of allergies, on the basis that cross-sensitization is commonly observed. In the longer term, a well-validated animal model of food allergy is needed.

Host Metabolism

As indicated in the previous section, many of the genes now being introduced into experimental crops to provide insect resistance depend for their action on disrupting the digestive function of the pest. Some of the enzyme inhibitors and lectins being considered produce similar toxic effects in mammals and many of the processes, commercial or domestic, currently used to prepare foods and animal feeds are designed to degrade or remove anti-nutritional factors of this type. Plant enzyme inhibitors belong to a general class of defence proteins, some of which are known to be cross-reacting allergens.[61] Activation of plant defence proteins has been shown in *Brassica rapa* to considerably increase the number and intensity of bands visible in IgE immunoblotting recognized by the pooled sera of patients allergic to latex.[62] Introduction of enzyme inhibitors or overexpression of existing defence proteins might induce similar effects. Lectins and some enzyme inhibitors also are known to be highly resistant to degradation in the digestive tract and can be absorbed largely intact and appear in the blood stream. Where this occurs, many aspects of metabolism can be affected, including the induction of immune responses other than IgE and changes to the endocrine system. Validated methods to screen for effects on digestive enzymes and for more systemic effects need to be identified and developed before constructs of this type reach the point of seeking commercial release.

Inadvertent Effects

The various analytical parameters selected to examine whether a crop is substantially equivalent to other varieties of the same crop are unlikely to detect inadvertent effects other than those which directly affect the concentration of a known toxicant or which substantially disrupt plant function. Small changes to the gross composition are unlikely to be detected against the background variability introduced by environmental factors. If establishing the absence of inadvertent effects is seen as important, then it would be better served by the use of techniques which make no assumptions but which attempt to measure the construct as a whole. Several relatively new technical approaches would allow a more discerning analysis. These might include:

1. DNA array methods able to detect differences in total mRNA induced by other genes as well as the transcribed gene and to provide evidence of gene silencing. It should be recognized, however, that often there is a poor

---

[61] M. Mena, R. Sanchez-Monge, L. Gomez, G. Salcedo and P. Carbonero, *Plant Mol. Biol.*, 1992, **20**, 451.

[62] A.-R. Hänninen, J. H. Mikkola, N. Kalkkinen, K. Turjanmaa, T. Reunala and T. Palosuo, *J. Allergy Clin. Immunol.*, 1999, **104**, 194.

correlation between the concentration of the messenger and the amount of product expressed.

2. Proteomics to detect differences in total protein expression. This would allay concern about whether the introduced genetic material had disrupted expression of an existing gene or led to different protein structures because of modified post-transcriptional changes.

3. Metabolic profiling techniques could lead to a clear documentation of amplified or suppressed metabolic pathways not considered in current testing systems.[63] Analyses of metabolic pathways are clearly important when an introduced gene is known to code for an enzyme involved in the production of plant secondary metabolites.

Although such techniques are of considerable potential, their application to safety testing is novel. As yet, there is little experimental evidence available to establish the natural variation to be expected, whether results are capable of being interpreted and whether unexpected differences can be detected. An ability to detect inadvertent changes will become of increasing importance as the complexity of genetic modifications increase. Already examples of conventional crosses between GM plants expressing different introduced traits to provide hybrids containing both have sought market approval. The yellow rice provides an example in which multiple introductions were made to establish a novel pathway.[46] It has to be assumed that, in future, parent lines are more likely than not to be genetically modified, initially to provide a background of pest resistance or herbicide tolerance into which genes controlling quality traits could be introduced. From a safety viewpoint this process of 'gene stacking' (gene pyramids) introduces the question of interactions between introduced traits and increases the likelihood of inadvertent effects. It also has been argued that tDNA is likely to be randomly introduced into retrotransposons, mobile genetic elements which make up about 50% of nuclear DNA in plants, and that altered spatial and temporal expression may occur and that transposition is possible.[64]

7 Conclusions

A first and essential step in any safety evaluation is the identification of possible hazards, however remote or implausible. It is the subsequent considerations in risk analysis, that take account of the likelihood of occurrence and severity of effect, which introduce an element of proportion into the evaluation process. Unfortunately, much of the debate surrounding the introduction of GM crops and their safety for humans has stopped at the level of hazard identification. There is, in fact, no evidence that suggests that the present generation of GM crops pose any risk to human health, although it would be difficult to be quite so categorical about risks to the wider environment. This is no cause for complacency, however. Assessments made to date have been developed with the experience gained from a very limited number of constructs and from a perspective which assumes equivalence to conventional crop varieties. Even this

[63] H. P. J. M. Noteborn, A. Lommen, R. C. van der Jagt and J. M. Weseman, *J. Biotechnol.*, 2000, **77**, 103.
[64] B. Jank and A. Hasleberger, *Trends Biotechnol.*, 2000, **18**, 326.

limited experience has triggered research that, while not casting doubt on the evaluations completed, has challenged a number of beliefs. Typical of this was evidence of the survival and uptake of DNA from the gut,[65,66] which was previously assumed not to happen. Far greater challenges to risk assessment will be made in the future when comparisons to existing conventional varieties may not provide a point of reference. Risk assessment methods must stay abreast of developments in GM technology. Uncertainty and concern will arise only when the means of addressing safety issues are not adequate to the deal with the latest modifications seeking authorization and release.

8   Acknowledgement

The Author acknowledges the support of the Scottish Executive Rural Affairs Department (SERAD, formally the Agricultural, Environment and Fisheries Department of the Scottish Office).

---

[65] R. Schubbert, C. Lettmann and W. Doerfler, *Mol. Gen. Genet.*, 1994, **242**, 495.
[66] R. Schubbert, D. Renz, B. Schmitz and W. Doerfler, *Proc. Natl. Acad. Sci. USA*, 1997, **94**, 961.

Safety and Quality Research Priorities in the Food Industry

GEOFF ANDREWS, ALASTAIR PENMAN AND CHRIS HART

1 Introduction

Food safety and food quality is a vast subject area. To address it fully would require an analysis of the complete food chain from seed or livestock genotype, through primary agriculture, primary, secondary and tertiary processing, formulation, packaging, distribution, retailing, domestic storage and finally consumption. Out-of-home consumption is steadily rising: already 35% of all meals in the UK are consumed out of the home. The year 1999 saw this figure in the USA exceed 50% for the first time. Catering and food service play an increasingly important part in our experience of food quality and food safety.

This article does not set out to identify all the issues impacting food safety or food quality. Rather, it attempts to identify those areas that require further scientific endeavour so that the foods and drinks available to the UK consumer will be safer and of higher quality in future years.

Food safety is not limited to microbiological safety. As recent history has demonstrated with BSE and vCJD, anaphylactic shock from eating peanuts, dioxins entering the human food chain via animal feedstuffs, benzene in mineral water and glass fragments in baby food, food safety also includes chemical contamination and foreign bodies. Prions, the cause of BSE and vCJD, are an entirely new source of food-borne disease. Food-borne viruses are becoming recognized as significant to public health. As the examples of benzene and dioxins demonstrate, food safety is not necessarily about real risk to public health, but also about perceived risk.

For this article, food quality will be limited to a discussion of the organoleptic quality enjoyed by the consumer. However, this is a complex area which is dependent on the quality of the raw materials, and on subsequent processing and storage. Organoleptic quality has both whole food components (does the loaf look good) and molecular components (is the starch in the right crystalline state).

Issues in Environmental Science and Technology No. 15
Food Safety and Food Quality

In many ways the story of the food industry is one of great progress and success. The industry offers the consumer an ever-widening choice of products with increased convenience and great value for money. The percentage UK household income spent on food has fallen consistently as the quality of the foods has increased. However, the industry is partly a victim of its own success: the consumer has come to expect and demand quality and value to improve constantly and at the same time is increasingly intolerant of failures in either safety or quality. The industry must continue to harness good science to continue the journey towards zero defect products that are still great value and great quality.

This article reflects the views of the Food and Drink Federation Food Research Policy Group. The Group, consisting of the senior technical staff from most of the UK's largest food manufacturers, seeks to establish consensus within the food manufacturing industry on food research priorities and to inform the direction of publicly funded research to address the issues identified. Papers on research priorities in Food Safety, Food & Drink quality, Diet & Nutrition and Sustainability are available from the FRPG Secretariat, FDF, Federation House, 6 Catherine Street, London, WC2B 5JJ.

2 Food Safety

Microbiological Safety of Foods

Incidence of Food Poisoning. PHLS statistics indicate that the rate of reported food poisoning incidents has declined slightly since 1997.[1] *Campylobacter* spp. are by far the most commonly reported gastrointestinal pathogens, with almost 55 000 cases reported in 1999, down from 58 000 in 1998. The incidence of salmonellosis has been steady at around 30 000 from 1989 to 1997, since when its incidence has almost halved. Since 1979, reported incidents of rotavirus have risen from zero to just under 15 000 in 1999 and the incidents of the protozoan cryptosporidium has climbed from zero in 1983 to around 4800 in 1999.

However, it is by no means certain that all the reported illnesses that are attributed to pathogens which can be food borne were actually derived from foodstuffs and not from other failures of hygiene. For example, *E. coli* 0157 is a food-borne pathogen but children have also been infected during visits to farms and both domestic and farm animals are recognized as a source of infection.[2,3] In addition, it is certain that not all food-borne infections are reported, so the real prevalence of food-borne disease in the UK is unclear.

The Infectious Intestinal Diseases survey[4,5] studied 70 GP practices that served 460 000 people and surveyed almost 10 000 patients. Data collection started in 1993 and concluded in 1996. The study showed that intestinal infection

[1] Anon., *CDR Weekly*, 2000, **10** (2), 1.

[2] M. Day, *New Sci.*, 1997, **154** (2086), 12.

[3] J.S. Wallace, in *Escherichia coli 0157 in Farm Animals*, ed. C.S. Stewart and H.J. Flint, CABI, Wallingford, 1999, pp. 195–223.

[4] J. G. Wheeler, D. Sethi, J. M. Cowden, P. G. Wall, L. C. Rodrigues, D. S. Tompkins, M. J. Hudson and P. J. Roderick *Br. Med. J.*, 1999, **318**, 1046.

[5] D. S. Tompkins, M. J. Hudson, H. R. Smith, R. P. Eglin, J. G. Wheeler, M. M. Brett, R. J. Owen, J. S. Brazier, P. Cumberland, V. King and P. E. Cook, *Communicable Dis. Public Health*, 1999, **2**, 108.

in the population is much higher than had been previously thought. Extrapolating from the results, during the period of the survey, each year, one in five persons would have suffered a gastrointestinal infection (9.4 million cases) but only one in six presented to a GP; there would have been 90 000 cases of salmonellosis (1.4 times the rate of incidence recorded by the GPs) and 350 000 cases of campylobacteriosis (2.1 times the rate of incidence recorded by the GPs); there were more cases caused by rotavirus A than by *Salmonella* spp.; only 1 in every 136 cases of infection were reported to the Communicable Disease Surveillance Centre; viruses were shown to be the most common cause of infection, particularly amongst children. However, the proportion of these infections for which food was the vector was not established.

These are important considerations. Intestinal infection is usually interpreted as food poisoning. The food industry will improve its performance and further reduce the possibility of pathogenic bacteria being carried in foodstuffs. However, with such under-reporting of intestinal infection, there is scope for the reported levels of intestinal disease to increase even when the actual incidence of food poisoning is falling. Although food may often be the vector, it is unknown what contribution other failures of hygiene contribute to the reported figures. A similar exercise to the Intestinal Diseases Survey that was extended to identify the source of infection would be enormously helpful to the manufacturing, catering and retailing industries, to statutory and regulatory bodies and to health professionals, to understand where efforts need to be focused to reduce the overall rate of food-borne infection.

However, the food industry continues to take food poisoning seriously and accepts that the bacteria most commonly associated with stomach upset, *Salmonella* spp. and *Campylobacter* spp., may well be food borne in the majority of cases.

Microbiological Risk Analysis. Hazard Analysis and Critical Control Points (HACCP) is now well established in the food manufacturing industry and is beginning to be applied to primary production. However, it is only one part of the risk analysis process defined as risk assessment, risk management and risk communication.[6] HACCP is a risk management tool, not a risk assessment tool. With the complexity of the food chain, and the need to provide the consumer with value, a microbiological risk assessment tool is required to ensure that the control procedures are commensurate with the risk identified. These points of risk will be different for different pathogens and may be different depending on the section of the population for which the final foodstuff is intended (children, elderly, hospitalized). HACCP has been widely implemented in the food industry because it is a simple tool to understand and use and it can be applied to any production process. The result of applying HACCP is a simple but powerful risk management model. However, the selection of the critical control points and the setting of control parameters are largely based upon judgement that does not assess relative risk. The development of a simple microbiological risk assessment tool is now needed to prioritize risk elimination or containment throughout the food

---

[6] J. L. Jouve, M. F. Stringer and A. C. Baird-Parker, *Food Safety Management Tools*, ILSI, Brussels, 1998.

chain and for use in conjunction with HACCP to ensure that appropriate levels of control are implemented.

Risk assessment is important for maintaining an industry that offers both safe and value-for-money foods. The precautionary principle is rightly invoked in the interest of protecting the consumer. It is applied in a situation where there is any doubt in consideration of consumer health: actions must favour protection of the consumer rather than doing nothing, taking a risk and hoping for the best. However, the precautionary principle, if overused, can have damaging effects. Product recalls and retrievals are expensive activities and we effectively pay for the cost of such events when we buy our foods. A too rapid invocation of the precautionary principle may result in unnecessary recalls and unnecessary costs that the consumer ultimately has to pay. Unnecessary recalls may result in a business folding, resulting in a reduction in consumer choice. Fear of recalls may be an unnecessary barrier to entry for new and innovative businesses. Overuse of the precautionary principle may also make it difficult for the retailer and statutory authorities to gain the attention of the consumer when there is a real and substantial threat to consumer health: it may be difficult to hear the shouts of warning over the clamour for attention to minor failures in product safety.

The precautionary principle is the correct approach to uncertain situations but needs to be informed by a robust assessment of risk. A protocol that has the support of manufacturers, retailers and statutory authorities is required.

Primary Agriculture. Even without a microbiological risk assessment tool, we know that quality assurance protocols encourage elimination of problems at source rather than dealing with the problem later in the supply chain. So for control of food-borne pathogens, the trail starts on the farm. *Salmonella* spp. and *Campylobacter* spp. are both most strongly associated with raw poultry. The poultry industry has made huge strides in the past decade to understand the source of *Salmonella* spp. which has involved a detailed understanding of the operation of feed mills, contamination routes through layeries and hatcheries, and cross-contamination in poultry houses. The infection rate of poultry in good quality operations can be expected to be 3% or less. However, once into the slaughterhouses and chicken processing operation, cross-contamination typically results in a 10-fold increase in the percentage of dressed carcasses that are *Salmonella* positive. This is a direct result of the scale and, therefore, mechanization of the process. With the UK consuming 15 million birds per week, mechanization and automation are unavoidable and, therefore, the probability of cross-contamination remains high.

Some processors claim to have eliminated *Salmonella* spp. from their flocks and abattoirs and are able to sell *Salmonella*-free chicken.[7] The lessons from such processors should be learnt and applied as widely as possible. In the short term, *Salmonella*-free chicken can be expected to be a premium product. In the medium term, we should expect that *Salmonella*-free chicken should become the standard for retail and manufacturing meats.

Although methods are being developed to reduce carcass contamination, such

---

[7] H. Armstrong, *Meat Int.*, 1997, **7** (2), 16.

as whole carcass steam pasteurization,[8] the pressure must remain on the primary producers to completely eliminate *Salmonella* spp. from the flock. That this is possible is demonstrated by some Danish producers who claim to have *Salmonella*-free flocks.[9] However, the simultaneous pressure to reduce or eliminate the use of antibiotics in poultry production[10] will make this objective even harder to achieve.

The success in reducing *Salmonella* spp. in the poultry production chain has been in part due to the ability to accurately identify the serotype. This has enabled producers to trace the source of the *Salmonella* spp. back through the supply chain and eliminate it at source. No such opportunity is offered by *Campylobacter* spp., very fastidious microaerophiles that can be very difficult to culture. Typing the sub-species it also very difficult. Additionally, *Campylobacter* spp. have been shown to exist in a viable but non-culturable form so that failure to detect the organism cannot be taken as proof of its absence.[10] With *Campylobacter* spp. now the most commonly reported cause of food poisoning, new tools are required to identify and track *Campylobacter* spp. through the supply chain so that control and elimination protocols can be appropriately targeted.

In parallel, techniques for the elimination of this pathogen once in the poultry flock, that do not depend upon antibiotic treatment, are also required.

The incidence of *Campylobacter* spp. is also not understood. It is known that in two chicken sheds which appear to have been treated in an identical fashion and stocked with birds from the same hatcheries, one can be *Campylobacter*-free, the other heavily infected. Methods to trace the bacteria back through the chain may help to identify the reasons for these observed differences.

E. coli O157, the cause of haemorrhagic ureamic syndrome, and most notoriously linked with the outbreak traced to a butcher's shop in Lanarkshire,[11] is probably a recent mutation of this ubiquitous organism. Its source is almost certainly cattle, both beef and dairy, as exemplified by the Lanarkshire butcher and more recently by a small dairy in Whitehead, Cumbria.[12] It has to be accepted that where there are livestock there also faeces. Dairy farmers have significantly increased their understanding of udder hygiene prior to milking that has contributed to the decline in total viable count in raw milk in the past 15 years. Clearly, however, some risk of contamination of the milk still exists, although this should be controlled by proper pasteurization. In the case of the Whitehead milk producer processor, the infection was caused both by contamination of the raw milk and the improper control of the pasteurizer.

In the slaughterhouse, the animal can be cleaned before slaughter but the dead animal carries a reservoir of infection in its intestinal tract into the first stages of the butchery. It is almost inevitable that the gut contents will occasionally come into contact with the muscle or the slaughterhouse equipment, making

8   A. L. Nutsch, R. K. Phebus, M. J. Reimann, D. E. Schafer, R. C. Wilson, J. D. Leising and C. L. Kastner, *Int. Food Hyg.*, 1996, **7** (6), 21.
9   Advisory Committee on the Microbiological Safety of Food, *Report on Microbiological Antibiotic Resistance in Relation to Food Safety*, HMSO, Norwich, 1999.
10 Advisory Committee on the Microbiological Safety of Food, *Interim Report on Campylobacter*, HMSO, London, 1993.
11 T. H. Pennington, in *Farms and E. coli O157 in Livestock*, Stationery Office, Edinburgh, 1997, p. 15–17.
12 Anon., *CDR Weekly*, 1999, **9** (12), 108.

contamination of the carcass almost unavoidable. The greatest risk posed by infected meat is probably not in the major food processing operations, but in the potential for cross-contamination of cooked products in domestic and catering kitchens.

Although, as for poultry, whole-carcass steam pasteurization is being developed, the focus of effort should be at the source of contamination: the farm. The objective must be the elimination of *E. coli* O157 as a farm contaminant.

The objective of control of pathogens on farms raises the issue of the risk presented by conventional farming practice in the handling of animal faeces. When on pasture, cattle manure the land. If *E. coli* O157 persists in the environment, and the evidence suggest that it does,[2] then does this represent a source of infection for other cattle and for how long is there a risk of infection? When cattle are kept under cover, farmers have to remove large quantities of urine, faeces and contaminated bedding material. The solids are typically separated off in a lagoon and the slurry applied to the land and crops. There is evidence of *E. coli* O157 being found in salads grown on land that has been sprayed with animal waste slurry.[13,14] With a food-borne pathogen now in our beef and dairy herds, the practice of disposal of untreated animal wastes onto land must be called into question. *E. coli* O157 must either be eliminated from our herds, or the use of untreated animal waste must be controlled and regulated.

The use of farm animal waste goes beyond *E. coli* O157, as little is known about the persistence of pathogens, including viruses and parasites, in such waste and in the environment after spreading. Understanding the fate of animal-borne pathogens in the environment needs to be better understood before a microbiological risk assessment tool can be applied to direct the application of appropriate risk control measures.

The risk presented by the use of both treated and untreated human sewage sludge for agriculture were recognized in the development of the 'Guidelines for the Application of Sewage Sludge to Agricultural Land'.[15] This banned the use of untreated sewage sludge for almost all agricultural uses and banned the use of treated sewage sludge on all human food crops. Only sewage treated by methods which eliminate all pathogens can be widely used. If such measures are appropriate for human sewage sludge, but animals carry human pathogens and far more animal sludge is spread to land than human sludge, it would seem appropriate that similar measures are considered for animal wastes.

Viruses. Recent years have seen a new source of food poisoning to be recognized: food-borne viruses. The Infectious Intestinal Diseases Survey[4] showed that viral infection is a common cause of intestinal disease, with the incidence of rotavirus A and small round structured viruses exceeding the rate of incidence of *Salmonella* spp. both in the community and as reported to general practitioners. The significance of food as the vector for viruses is unknown, although there is clear evidence that viruses can be food borne. Viruses pose three

[13] E. Gutierrez, *Lancet*, 1997, **349**, 1156.

[14] C. Bell and A. Kyriakides, in *E. coli: a Practical Approach to the Organism and its Control in Foods*, Blackie, London, 1998, pp. 129–141.

[15] Anon., *Guidelines for the Application of Sewage Sludge to Agricultural Land*, ADAS, Guildford, 1999.

problems to the food industry: effective control of viral contamination, measurement of levels of viral contamination (much more difficult than measuring bacteria) and managing staff infected with some types of food-borne virus. The last point is an issue because of the rapid onset of the infection and the potential for a member of staff to be afflicted with projectile vomiting whilst engaged in a food handling environment.

Protozoan Parasites.   Cryptosporidium and Giardia are water-borne protozoa capable of causing food poisoning. Although their incidence is rare, they could enter a factory in the mains water supply. There is now a legal limit of 1 cryptosporidium oocyst per 10 litres and water companies must monitor their water to ensure that this standard is met. In the event of a protozoan contamination being detected, water companies typically issue a boiled water notice (BWN). Boiling water prior to consumption may be an acceptable, albeit inconvenient, method of preventing a protozoan infection in a domestic environment, but factories are not equipped to boil or further treat their fresh water supplies. Installing microfiltration in every factory would be a very expensive solution to managing the risk. The answer must be to eliminate them from the water supply. Although some of the risk factors that lead to contamination are understood, there is still a major gap in our understanding of how such a contamination occurs, the pattern of protozoan contamination during an incident, the best corrective measures, how to monitor background contamination levels and what level of contamination should trigger a response so that corrective action is taken long before a BWN is considered.

Another protozoan pathogen has been recently identified. Cyclospora infection[16] is associated with eating contaminated fruit and vegetables. The best documented outbreak[17] was attributed to fresh raspberries from Guatemala that had been sprayed with a contaminated pesticide solution; hundreds of victims were involved. The route of infection was never confirmed.

There is a general need to better understand the significance of protozoan infection and the role of food and to devise appropriate control mechanisms to prevent infection if their presence is detected.

Minimally Processed Foods.   When considering minimally processed foods, food quality impacts food safety. In the industry's effort to replicate the best standards of home-prepared and restaurant-prepared meals, we have attempted to minimize the heat processing that foodstuffs experience before being purchased by the consumer. For some foods (prepared salads, sandwiches), no heat processing has been applied to the key ingredients or to the finished product. For other products (chilled ready meals), ingredients and components are given the minimum heat treatment consistent with safety and shelf life and the assembled finished product receives no further heat treatment. The key tool for ensuring safety of such products is HACCP, which, as previously stated, would be strengthened by a microbiological risk assessment.

---

[16] S. C. Clarke and M. McIntyre, *Rev. Med. Microbiol.*, 1996, **7**, 143.
[17] R. McClelland, *Can. J. Med. Lab. Sci.*, 1997, **59**, 23.

Microbiological Testing. Although microbiological testing is routinely applied to such products, we know that the level of testing has no statistical validity (although over an extended period of time it can generate confidence in a process). What is the significance of a total viable count from 0.1 g sample of complex foodstuff, even when conducted in triplicate? It says nothing about the rest of the sample from which it was taken, let alone about the rest of the products made on that shift of which it is supposed to be representative. The temptation is to test more, but a statistically valid testing regime would make prepared foods unaffordable. The need is for risk-based sampling schemes that recognize the limitation of sampling and target where and when to sample. Indeed, the answer may be to re-direct resources away from microbiological screening of finished products and towards other control measures.

Current testing regimes are crude and do not recognize the condition of the bacteria in a foodstuff. Contaminating bacteria are transferred from a foodstuff to a liquid or semi-liquid environment, designed to provide nutrients for optimal growth and then incubated at optimal growth temperatures. The reality is that the bacteria may be sub-lethally damaged, in an environment that provides poor nutrition which is maintained at a sub-optimal growth temperature. In solid foods, the bacteria may find themselves immobilized in the food structure in which migration of nutrients to the bacteria and migration of growth limiting metabolites away from the bacteria are severely limited. In such cases, growth models derived from planktonic environments may seriously overestimate the real multiplication of bacteria.

The research need is clear: what are the growth models that most closely reflect the actual proliferation of bacteria in foodstuffs. Should we continue to use the 'worst case' model of planktonic (*i.e.* in broth or on a plate) growth? Does a more accurate model enable the industry to offer minimally processed foods or a longer shelf life at no additional risk to the consumer?

The additional complexity is this area is the difference between the presence of pathogenic bacteria and their infectivity and how this is modified by growth in sub-optimal conditions in or on a foodstuff. Which are the most susceptible social groups upon which maximum levels of contamination should be based? Is infectivity or virulence affected by growth conditions? Are pathogenic bacteria more or less virulent as a consequence of growth in a food medium compared to a microbiological culture?

Chemical Safety

Chemical safety of our foods is a more complex area than microbiological safety.

Perception of Risk. For most people, infection by pathogenic bacteria usually results in a unpleasant but short-lived experience within a day or so of ingestion of the contaminated food. No permanent damage is normally done. The food responsible can often be identified. Most of us have experienced it at some time or other and, even if we felt we were about to die, we survived it. Bacterial food poisoning for the majority of the population holds no particular fear. However, for parents of young children, for the elderly and infirm and for the immunocompromised, there is more reason to be careful to avoid food-borne infections.

In contrast, the effects of chemical contamination are less evident. Although some may trigger a swift reaction, the concern is normally that a one-off exposure will cause a disease condition long after the exposure (*e.g.* carcinogens) or that chronic exposure will produce a slow irreversible degeneration (*e.g.* lead, mercury). Such disease conditions are more readily seen as permanently impairing or fatal. Also, we are continually improving our understanding of the potential of environmental contaminants to impair health. Organophosphates have become increasingly restricted as their toxicity has been recognized. 3-Monochloropropanediol (3-MCPD), a carcinogen has been found to be a breakdown product in some food processes, and potentially present in a wide range of foodstuffs. Pesticide and herbicide residues are becoming ubiquitous in both the developed and developing agricultural economies and, with improving sensitivity of analytical techniques, can be found in many products, albeit at levels below 1 ppb.

The research in this area, therefore, not only needs to address the absolute risk of chemical contaminants in the food chain, but it must also satisfy the risk as perceived by the consumer and provide for the uncertainties caused by chronic exposure to very low levels.

Veterinary Residues & GMOs.   Recent developments have brought two new contaminants to the public and regulators' attention: veterinary residues and genetically modified organisms. The reaction has been both from a food safety and an ethical standpoint, but we will address only the food safety component.

The veterinary residues that give rise to concern are either hormones or antibiotics. Research and environmental surveillance has demonstrated the persistence of oestrogens in our rivers[18] and pointed to the effect of oestrogens and oestrogen mimics affecting the sexual development of fish.[19] By extension, there is concern that hormone-treated dairy herds and beef cattle could have a damaging effect on the human population. We do not know if this is a real risk, or how great this risk is or whether different consumers are more liable to have an adverse reaction.

Antibiotics have been widely used in livestock production to treat disease conditions, prophylactically to prevent spread of a disease condition in a flock or herd, or to suppress pathogenic bacteria so as to improve weight gain in the animals. The Advisory Committee on the Microbiological Safety of Foods recently recommended a significant reduction in the use of antibiotics on the farm[9] because of two possibilities: the development of antibiotic-resistant pathogenic bacteria in the livestock, which could result in a hard-to-treat infection of the consumer, and the ingestion of antibiotics causing the selection of antibiotic-resistant bacteria in the gut. There appears to be a conflict between the desire to suppress pathogenic bacteria in the herd or flock and the long-term effects of this strategy on the development of antibiotic-resistant bacteria. The research needs in this area are the development of pathogen control strategies

---

[18] Environment Agency Research & Dissemination Centre, *ENDS Report*, 1999, p. 15.
[19] Anon., *ENDS Report*, 1995, p. 4.

that are not antibiotic dependent, better methods of detecting and isolating diseased animals before the herd or flock is infected, and a better understanding of how ingested antibiotics affect our own gut flora.

Although there is no evidence that genetically modified tomatoes, soy beans or maize pose any public health risk, nevertheless, there remain concerns about their long-term effects. As a new technology, these are uncharted waters for both the suppliers of these crops, for the industry and for the consumer. Although much research has already been conducted on the potential impact on health (allergenicity of new proteins, consequences of using antibiotic marker genes, potential for transfer of transgenic DNA into other cells), continuing studies are required as new GMOs are brought towards commercial production and on the long-term health outcomes of growth and ingestion of GMOs. The research programme of the BBSRC's Institute of Food Research[20] into stable and more predictable ways of modifying a genome is very relevant for the long-term exploitation of this technology.

Other Contaminants.   There is potential for food to be contaminated at any point along its sometimes lengthy and tortuous supply chain. Therefore, environmental contaminants can become food contaminants and chemicals used for packaging can find their way into foods. As the consumer becomes more aware of the potential for contamination and more insistent on elimination of risk, the industry has to be more alert to potential sources. In parallel, our analytical techniques are becoming more sensitive and we are finding chemical traces in our foods that had not previously been identified. 3-MCPD is an example: as an artefact of heating protein in the presence of lipids, it must have been in our foods for many years. More recently, it has been identified as a carcinogen and analytical methodologies have been developed to enable its measurement in foods at the ppb level.

The contamination of Belgian animal feed with dioxins, in the first half of 1999, was inexcusable and a glaring example of malpractice. Nevertheless, the levels of dioxins entering the food chain, although detectable, were too low to be a risk to public health. The precautionary principle was invoked which required contaminated feed and livestock to be segregated because of the potential for unacceptably high levels of dioxins entering the human food chain.

Phyto-oestrogens are a newly recognized potential threat to food safety.[21] Although they may be an intrinsic component of food ingredients, and not a contaminant, they may have a deleterious affect on health if consumed in sufficient quantity over a sufficient period of time. However, we do not yet know enough of their true impact on human health either in the short term or following chronic ingestion.

Mycotoxins and ochratoxins are metabolites from certain fungi. The threat of mycotoxin contamination from, for example, mouldy peanuts, was recognized two decades ago. The industry took the sensible precaution of implementing quality systems that reduce the possibility of mould growth and reject materials

[20] Anon., *The Safety and Impact of Genetically Modified Organisms* in *BBSRC Business*, BBSRC, Swindon, October 1998, pp. 3–5.
[21] S. Fritsche, *Lipid Technol.*, 1999, **11** (4), 82.

that have been, or might have been, contaminated. However, it is impossible to be sure that not one single nut in a consignment of many tonnes has supported fungal growth. Analytical techniques are sufficiently sensitive to detect mycotoxins when not even a detailed visual inspection would have identified contamination. In such cases, the presence of mycotoxins at low levels should not be taken as an indication of a potential risk to food safety but as a prompt to ensure that quality systems are working as designed.

In each of these cases, the industry requires an objective measure of real risk and decision-making criteria which have wide acceptance that allow for an appropriate response to an unforeseen event.

Foreign Bodies

As for chemical contaminants, physical contaminants may pose both a safety risk and a risk of perceived degradation of quality. A foreign body rarely impacts food safety. A piece of glass may cause a cut mouth or tongue but is unlikely then to be ingested. Foreign bodies will affect only a small number of consumers, in contrast to the potential for microbiological and chemical contamination which could affect whole populations. However, when glass is found in baby food or a hypodermic needle in a soft drinks can, the issue becomes centred on food safety. If the contaminant is peanut residue in a product that is otherwise nut free, the potential for mortality is present.

The more typical response to a foreign body contamination is one of revulsion (an insect in a prepared salad) or dissatisfaction (a piece of wood in a fruit pie). The economic justification for addressing these as serious issues arises from the need to protect a brand name, whether that is the brand of the manufacturer or the retailer. Rather than the brand being damaged by a spate of consumer complaints, the reaction is usually to withdraw the product that might be affected. If the foreign body and its source can be identified and traced to a source in the supply chain, the depth and breadth of a withdrawal can be limited. Where such certainty is lacking, withdrawals may be more widespread and the manufacturer may be required to implement expensive surveillance routines to prevent re-occurrence.

The challenge for the industry is to find more sensitive methods of checking ingredients, semi-finished goods and finished product for contamination. Can metal detectors reliably find ever smaller pieces of metal, even when the contaminant is non-ferrous and the product is in an aluminium tray? Can non-metallic contaminants such as chicken bones, or splinters of wood just a millimetre in length, be detected in a complex food product by X-ray or γ-ray instruments? Can the instruments operate fast enough to match a beverage canning line operating at 2000 cans/minute?

Methyl bromide is a very effective fumigant for elimination of insects from manufacturing environments. The population of such insects as flour moth and cocoa moth are normally controlled by the periodic fumigation of a factory. The gas rapidly disperses to levels that are not toxic to humans, leaves no harmful residues and does not taint foodstuffs. Unfortunately, it is strongly ozone depleting and it will soon be illegal to use it for fumigation purposes. An alternative fumigant that has all the useful properties of methyl bromide but does not deplete ozone has not been identified. It is urgently needed. The potential for a significant increase in consumer complaints due to insect parts in foods exists.

Risk communication

The key requirement for managing incidents of food contamination is a risk analysis tool that has the support of the whole food chain, including consumers and regulatory authorities. There must always be a bias in favour of consumer protection, but draconian measures in response to a vanishingly small risk will, in the long term, be damaging to the industry and the consumer. One of the essential outputs of any risk analysis must be risk communication in a language that consumers and the media can understand, interpret and act upon. The language must quantify the risk in an unambiguous way so that consumers may exercise informed choice. This is a much needed area of research: the vision of a simple, universally understood, universally applied, symbol-based risk communication scale from 'no known risk' to 'do not consume' is very appealing.

3 Food Quality

Quality and Manufacturing

The quality of a consumer product has two very different components. The manufacturer may believe that he is increasing the quality of a product by adding more meat to a meat pie or using fresh not frozen vegetables in formulating a quiche. However, is the consumer able to discriminate between the standard and 'quality improved' product, and are they motivated by the proposition of such a quality improvement? This is the essence of the challenge when conducting sensory analysis: what are the sensory characteristics which translate into a perception of quality by the consumer that may increase propensity to purchase? Although some work has been done in this area,[22,23] it is still poorly understood. Ideally what is needed is a framework for product developers that does not need to be continually verified by expensive consumer research.

Further back along the food chain, manufacturers want to understand how to select raw materials that will impart the desired quality characteristics to the finished product. Although some raw material characteristics can be linked (sugar content of potatoes for frying, moisture content of wheat for milling), there is a general lack of understanding of how subtle differences in raw materials can profoundly affect product quality.

Despite decades of research, there is still much to be learnt about the impact of subtle changes in cereal chemistry upon brewing and distilling processes.[24] That such subtle changes have an effect is certain but the molecular basis for this is far from clear. Understanding what is going on at a molecular level would allow

[22] D. A. Booth, *J. Market Res. Soc.*, 1998, **30**, 127.

[23] D. A. Booth, *Br. Food J.*, 1992, **93**, (9), 17.

[24] M. Moll and J. J. De Blauwe, in *Beers and Coolers: Definition, Manufacture, Composition*, ed. M. Moll and J. J. De Blauwe, Intercept, Andover, 1994, pp. 21–56

either for a more informed selection of raw materials or for an intervention in the process that was dependent on the raw material feed quality.

Other examples include understanding what it is that causes milk functionality to change during the spring flush.[25,26] Despite little change in the gross composition, the functionality can change dramatically. Why? What should we measure in order to adjust our processes?

In the produce arena, we want to be able to ask what biochemical or physiological markers should we look for when selecting a cauliflower that will retain physical structure when cooked and not fall apart?

The need in these areas is for tools and techniques that allow the food industry to move away from empiricism in selection of raw materials and towards knowledge-based decision making.

When we want tender steak, is there a better way of selecting it other than specifying which cut or measuring a textural characteristic? Is there a biochemical marker that translates into tenderness? Can we have meat that is both tender and flavoursome? Why does farmed salmon taste different to wild salmon and what do we have to do to get the wild flavour into farmed salmon? If chicken meat from a 12-month-old hen is more flavoursome than from a 42-day-old chicken, what are the biochemical markers upon which we could select for breeding?

Loss of quality

Just as important as selecting for food quality is the prevention of the loss of quality. The most pervasive problems are probably control of staling, and prevention of moisture migration.

Staling as a process is well understood and much progress has been made in understanding[27] how to delay the staling process in wheat flours. However, cooked rice stales and becomes hard within 24 hours of cooking; arresting the staling process would allow cooked rice of good quality and practicable shelf life to be offered. Much remains to be understood to control staling so that it can either be stopped when it has reached a desired level or completely inhibited.

Moisture migration limits the shelf life of many products[28] where a high moisture region is in contact with a low moisture region. The manufacturer is unable to give the consumer the experience of a fresh baked product. This is exemplified in pork pies. The pastry is designed to be hard and brittle to contrast with the moist and malleable meat content. Around the meat, a high moisture jelly is injected after baking. Within a few days, moisture migrates from the jelly into the pastry and the case becomes soft, losing both flavour and textural contrast with the filling.

25 A. M. O'Keefe, *Irish J. Food Sci. Technol.*, 1984, **8** (1), 27.
26 A. W. Nichol, T. J. Harden, C. R. Dass, L. Angel and J. P. Louis, *Aust. J. Dairy Technol.*, 1995, **50** (2), 41.
27 D. R. Linebackl, *Cereals International: Proceedings of an International Conference, Brisbane, September 1991*, RACI, Parkville, 1991, pp. 64–69.
28 P. Chinachoti, in *Food Storage Stability*, ed. I. A. Taub and R. P. Singh, CRC Press, Boca Raton, 1998, pp. 245–67.

Tailoring Raw Materials for the Market

Both conventional breeding and more particularly the promise of direct genetic modification of genotype offer the potential for changing the marketing dynamic of fresh produce, milk and meat.

Fruit and vegetables marketed to the consumer are still predominantly producer led: what is offered is what can be grown. Quality criteria reflect, not what is desired, but the limits within which a product is acceptable or tolerated. By using advanced breeding techniques, the food industry would want the potential to match a fruit or vegetable to the consumers' preference and the consumers' mode of using the product. This would be true of fresh produce for manufacture, too. The dream is to be able to specify the ideal quality criteria of a fruit or vegetable and know that a grower is able to match the characteristics accurately and consistently. The desirable characteristics would include flavour, sweetness, texture, nutrient content and durability in storage. No doubt the growers would want to specify such characteristics as yield, drought tolerance, insect resistance and ease of harvest.

Similar scenarios can be imagined for milk and meat. As the market demand for butter fat, casein and whey protein moves, one or other component is in surplus, another in dearth. Although some changes in gross composition have been achieved through conventional breeding and feed regimes, could milk composition be more closely tailored to market demand by a better understanding of feed conversion or by altering the genotype? Would this open up new opportunities for tailoring not just the quantity of caseins or whey proteins but also their functionality?

In the livestock market, the demands for the different cuts of meat must be managed. Chickens yield a fairly steady ratio of leg to wing to breast meat. However, the market does not necessarily reflect this balance. Beef muscle is selected from different parts of the animal according to the desire for leanness, tenderness and flavour, but the market for hind and fore quarter does not necessarily reflect the fact that for every forequarter there is exactly one hindquarter. The aspiration is not for a chicken with three legs and two wings, or a pig with only hind legs. So, can we breed chickens so that the eating quality of the leg is matched by that of the breast meat? Can we breed steers so that the fat/lean ratio of a beef forequarter matches the market demand rather than the market having to accept what is available?

Processing

Being able to specify the perfect raw material for processing pre-supposes that we know what happens during processing in sufficient detail such that we can specify the key raw material characteristics. Although much work has been conducted on various model systems to elucidate the fundamental changes that occur during a process, foodstuffs are often complex in composition, multiphase and non-Newtonian, which makes for very complex modelling. Nevertheless, there is a need for mathematical models that can inform the process engineer and food technologist about mass transfer and heat transfer, from which much can be inferred about the driving forces for changes at the molecular level.

Colour and flavour are the dominant quality characteristics that we seek to impart in heat processing. These characteristics are typically the result of the formation of a range of compounds from such recognized pathways as the Maillard reaction. The Maillard reaction has been the subject of extensive research and much is understood[29] about the chemical pathways and the reaction kinetics. Again, however, much of this has been conducted in model or simple systems. The challenge is to understand such pathways in complex food systems in such a way that progress down the pathway can be manipulated to achieve the desired finished product quality.[30]

Control of many processes tends to be empirical: trial and error will establish the process parameters required to deliver a particular sensory characteristic. With variable raw materials, however, processes may need to be tweaked to achieve a consistent finished product. Control systems are based around measurement and control of physical variables: moisture level, oven temperature, rheological properties. The consumer, however, experiences organoleptic quality. The challenge for the manufacturer is thus two-fold: can we link organoleptic quality to physical parameters such that varying these parameters has a predictable impact on organoleptic quality? Or can we measure organoleptic quality directly with on-line instrumentation, ideally with feedback control loops to maintain the organoleptic quality within specified limits? This is a major challenge to the sensory analysts to link organoleptic evaluation with physical processes so that the process engineer can design the control systems. For on-line instrumentation, the sensory analyst needs first to correlate an organoleptic quality to a physical (colour, rheology) or chemical parameter (level of a particular volatile compound). The instrument engineer then needs to design a means of measuring this in real time on-line. Finally, the influence of process parameters needs to be correlated with the organoleptic parameter so that a control loop can be designed.

The Meaning of Fresh

The word 'fresh' in the marketing of foodstuffs has become something of a mantra in the past decade. However, what the manufacturer and retailer mean by fresh and what the consumer understands are sometimes very different. We understand that fresh bread has been baked within the last 48 hours and fresh milk came from the cow within a similar period of time. So can fruit juice (the raw material could be weeks old), just because it has been pasteurized and requires chill chain distribution (rather than high temperature pasteurized and aseptically packed) be justifiably called 'fresh'? Also, 'fresh' fruit juice could have a 30-day best-before life: is it still 'fresh' after 30 days? 'Fresh' fruit and vegetables could be several days old at the point of sale.

'Fresh' has three connotations to the consumer: the first is about how recently it was derived from its source, the second is about minimal processing and the

[29] C. R. Lerici and M. C. Nicoli, in *Maillard Reaction in Food: Proceedings of a Round Table*, ed. C. R. Lerici, University of Udine, Udine, 1996, pp. 11–20.
[30] A. Huyghues-Despointes and V. A. Yaylayan, *Food Chem.*, 1994, **51**, 109–117.

G. Andrews, A. Penman and C. Hart

third about its eating quality. Pasteurized milk could be less than 24 hours from the cow when placed on sale. UHT milk when processed is no older than pasteurized milk but no-one would describe UHT milk, even on the day of processing, as fresh. In a supermarket, fresh fish means raw fish: it was probably caught several days ago and has been frozen.

The consumer does not experience age of product or type of process: the consumer will buy fresh foods on the basis of a superior eating quality, often expect to pay a premium and accept a short shelf-life. For the industry to offer both better value and greater convenience in the fresh foods market, we need to better understand what the consumer experiences as fresh food and how they discriminate 'fresh' from 'processed' foods. Armed with this information we would be better able to offer the fresh food eating experience that is both good value and convenient.

Date Coding

One of the ways that we communicate freshness to the consumer is by the best-before date on packaging. This is a very blunt and imperfect instrument. Chilled foods that have been temperature abused in the distribution chain may not be palatable or wholesome at their best-before date. Conversely, chilled products that have been maintained at below 5 °C may well be very acceptable long after the best-before date has passed. Many consumers misunderstand the significance of the best-before date and the product becomes a time bomb in the fridge: perfectly acceptable to consume up to midnight of the final day of the date code and potentially dangerous as the clock moves past 00:00.

Packaging

The manufacturer wants to delight their customers every time: we do not want consumers to eat temperature-abused product and risk losing a customer, but we do want the consumer to know when the product is still of the eatng quality to which we aspire. The challenge is then to find a much better method of communicating the quality of a product other than by a best-before date. Although time–temperature indicators have been developed and evaluated, they too are crude instruments: a short exposure to a high temperature may cause the time–temperature indicator to register temperature abuse and non-saleability of the product, but the product itself may be entirely unaffected. Identifying the quality characteristic or characteristics that a consumer expects in fresh food and then developing a low cost means of measuring this and clearly communicating it to the customer is the target, the so called 'smart packaging'.[31]

Consumer Choice

Finally, the link between quality and consumer choice is far from understood.[32] How does the consumers' understanding of quality match up to the industry's

[31] R. Ahvenainen and E. Hurme, *Food Addit. Contamin.*, 1997, **14**, 753.
[32] R. A. Dreiedonks, A. J. Eykelboom and J. B. F. Morel *Voedingsmiddelentechnologie*, 1987, **20** (26), 11.

understanding? How strong a factor is the consumers' perception of quality in comparison to other factors such as price and brand in influencing consumer choice? A better understanding of consumers will enable the industry to meet their needs and offer them a wide range of nutritious, delicious, attractive and good-value products.

Biosolid Recycling and Food Safety Issues

JIM WRIGHT

1 Summary

Sewage sludge amounts to typically 1% of the organic matter spread on agricultural land, the remainder being agricultural wastes and exempt industrial wastes, both of which, at the time of writing, are not subject to regulatory control.

The volume of sewage sludge produced is increasing every year and there is a growing requirement to expand the beneficial practice of recycling treated sewage sludge (biosolids) to agricultural land. However, various stakeholders, including food retailers in the UK, have expressed concerns over food safety issues in relation to the practice.

In both the USA and the UK there are complex regulatory systems for controlling recycling in order to protect: human, animal and plant health; ground and surface water quality; long-term soil quality; and soil biodiversity. These systems have been developed, using risk assessment techniques when possible, based on the significant reduction of pathogens in biosolids and ensuring there are physical barriers to avoid direct contact with pathogens.

The system for managing recycling in the UK, developed in 1989 in the form of a Code of Practice, has not adequately addressed new and emerging hazards and in particular, *E. coli* O157:H7, viruses generally, *Campylobacter*, *Listeria* spp., *Salmonella typhimurium* phage type DT 104, *Cryptosporidium*, *Giardia* and the BSE agent. In response, the UK water industry has undertaken a number of initiatives including the development of the Safe Sludge Matrix, a voluntary code of practice that is more stringent than the present legislation. In addition, the industry has commissioned extensive research into the fate of pathogens during the sewage sludge treatment and biosolid recycling process. During the development of the Safe Sludge Matrix, the UK water industry gave an undertaking to develop a quality assurance system for biosolid recycling.

The food industry normally applies Hazard Analysis and Critical Control Point (HACCP) to control food safety and quality issues. Through the application of HACCP to biosolid recycling, a number of Critical Control Points (CCPs) have been identified, including a relatively small number of regulatory

---

Issues in Environmental Science and Technology No. 15
Food Safety and Food Quality

CCPs concerned with pathogens and the presence of potentially toxic elements in the sludge. In addition, a number of quality related CCPs were identified. Human health safety issues can be controlled through the application of HACCP to the recycling process. Also, the HACCP process contributes to enhancing biosolid product quality.

2 Introduction

Organic Wastes other than Sewage Sludge and Recycling to Agricultural Land

Sewage sludge typically accounts for some 1% of the organic waste spread on land, the vast majority (96%) being agricultural waste and exempt industrial wastes (3%).[1] The latter are wastes, often referred to as industrial wastes, exempt from licensing under the Waste Management Licensing Regulations, 1994, and include food and drink wastes, paper waste, blood and guts from abattoirs, waste soils, dredgings and septic tank sludges and sludges from biological treatment plants.

Exempt wastes can arise from a wide variety of sources and therefore pathogen content is highly variable; however, exempt wastes can be variously categorized as low, medium or high risk, with abattoir wastes considered high risk.[2]

Agricultural wastes can be expected to have high levels of a number of the pathogens identified as being of concern in the recycling process. For example, it is estimated that *E. coli* O157: H7 is present in 2% of the national herd and that *E. coli* can remain viable for 70 days on grassland.[3] On a north Yorkshire farm, *E. coli* O157: H7 was found in the faeces of 16% of dairy cattle and 13% of beef cattle. In a north Yorkshire abattoir *E. coli* O157: H7 was found in the faeces of 1% of pigs and 2% of sheep.[2] *Cryptosporidium* spp. and *Salmonella* spp. are present in agricultural wastes, with the latter being able to survive for up to 100 days in slurry applied to grass.[4]

At the time of writing, agricultural wastes and exempt wastes are not subject to regulatory control, although it is expected that both will be brought under regulatory control in the near future. There are Guidelines,[5] intended to help farmers comply with the requirements of the EC Nitrate Directive (91/676/EEC), that are designed to reduce the likelihood of nitrate leaching. Specifically the Guidelines identify closed periods for the application of slurry, poultry manures and liquid digested sludge between 1st September and 1st November on grass or on ground to be sown with an autumn-sown crop and 1st August to 1st November to fields not in grass nor to be sown with an autumn-sown crop. These

[1] SEPA, *Strategic Review of Organic Waste Spread on Land*, 1998, p. 1.

[2] F. A. Nicholson. M. L. Hutchinson, K. A. Smith, C. W. Keevil, B. J. Chambers and A. Moore, *A Study on Farm Manure Applications to Agricultural Land and an Assessment of the Risks of Pathogen Transfer into the Food Chain*, ADAS Report FS2526, 2000, p. iii.

[3] Ref. 1, p. 29.

[4] E. G. Carrington, R. D. Davis and E. B. Pike. *Review of the Scientific Evidence Relating to the Controls on the Agricultural Use of Sewage Sludge. Part 1, The Evidence Underlying the 1989 Department of the Environment Code of Practice for Agricultural Use of Sludge and The Sludge (Use in Agriculture) Regulations*, Report DETR 4415/3, Water Research Centre, 1998, p. 29.

[5] MAFF, *Guidelines for Farmers in Nitrate Vulnerable Zones*, MAFF, London, 1998.

closed periods do not apply to farmyard manure (straw-based manure). The consequence of these closed periods is that farms must have sufficient storage capacity, typically 3 to 4 months, in order to satisfy the Guidelines.

Everyday farm management of agricultural wastes is variable, with practices ranging from immediate spreading of slurry through to storage for over 3 months prior to spreading. This latter condition generally results in low survival of pathogens. Solid manure storage for at least a month, where temperatures exceed 55 °C in the main body of the heap, is considered to ensure elimination of most pathogens. Recommendations for managing agricultural wastes include:[6]

- Storage of slurry for at least 1 month and preferably 3 months prior to application to land
- Solid manures should be stored for at least 3 months with turning and mixing to ensure 55 °C is achieved during composting prior to spreading
- Low trajectory slurry spreading should be employed to avoid inhalation of pathogen-laden aerosols
- Manures should not be applied directly to ready-to-eat crops and there should be at least 6 months between manure spreading and harvest
- There should be an interval of 6 months between harvesting and the time when the land was last used for grazing
- Manures should be applied to cut grassland rather than grazed pastures

Sewage Sludge

Sewage Treatment. Waste water is treated at a conventional treatment works to remove solids and reduce biochemical oxygen demand. This process is a controlled speeding up of the natural process of degradation and typically takes place in three main stages. Prior to these three main stages a preliminary stage consists of screening to remove litter and plastics, often followed by grit removal. The screened waste water then enters a primary treatment stage, commonly in the form of a settlement tank. Primary settlement typically removes 50% of the suspended solids, reduces biochemical oxygen demand by 20% and removes 50% of faecal coliforms. Usually a scraper at the bottom of the settlement tank collects the settled primary solids.

The secondary stage relies on micro-organisms that use waste water, from the primary treatment stage, as a source of nutrient. This process further reduces the biochemical oxygen demand and the numbers of organisms. When used efficiently, secondary treatment can remove up to 99% of faecal coliforms and 90% of indicator viruses.[7] The by-product of secondary treatment is microbiological biomass, or secondary sludge. A proportion of the secondary sludge is recycled to the secondary treatment process to maintain the viability of the process. The remaining sludge is consolidated with primary sludge.

Tertiary treatment is becoming increasingly common and is intended to achieve particularly high water quality standards for effluent prior to discharge.

---

[6] Ref. 3, pp. viii–x.
[7] Royal Commission on Environmental Pollution, *Sustainable Use of Soil, Nineteenth Report*, HMSO, London, 1996, pp. 81–104.

	Process	Description
Table 1 The seven treatment options in the code of practice	Sludge pasteurization	Minimum of 30 minutes at 70 °C or minimum of 4 hours at 55 °C (or appropriate intermediate conditions), followed in all cases by primary mesophilic anaerobic digestion
	Mesophilic anaerobic digestion	Mean retention period of at least 12 days primary digestion in the temperature range 35 ± 3 °C or of at least 20 days primary digestion in the temperature range 25 ± 3 °C followed in each case by a secondary stage which provides a mean retention period of at least 14 days
	Thermophilic aerobic digestion	Mean retention period of at least 7 days digestion. All sludge to be subject to a minimum of 55 °C for a period of at least 4 hours
	Composting (windrows or aerated piles)	The compost must be maintained at 40 °C for at least 5 days and 4 hours during this period at a minimum of 55 °C within the body of the pile, followed by a period of maturation adequate to ensure that the compost reaction process is substantially complete
	Lime stabilization of liquid sludge	Addition of lime to raise pH to greater than 12.00 and sufficient to ensure that the pH is not less than 12 for a minimum period of 2 hours. The sludge can then be used directly
	Liquid storage	Storage of untreated liquid sludge for a minimum period of 3 months
	Dewatering and storage	Conditioning of untreated sludge with lime or other coagulants followed by dewatering and storage of cake for a minimum period of 3 months. If sludge has been subject to primary mesophilic anaerobic digestion, storage to be for a minimum period of 14 days

There are a number of forms, including chemical precipitation, resulting in the generation of tertiary sludge.

Sludge Treatment. The sludge generated during waste water treatment is usually drawn off into holding tanks, where it is allowed to thicken through settlement. This settled sludge can then be further thickened using a variety of processes including belt presses, centrifuges and picket fence thickeners. This thickened sludge can be, depending on its final use, subject to a variety of treatments. If the sludge is to be used in agriculture there have been, since 1989, seven treatment options in the UK, as shown in Table 1.[8]

New waste water treatment works typically treat sludge by mesophilic anaerobic digestion followed by de-watering or thermal drying.

[8] Department of the Environment, *Code of Practice for Agricultural Use of Sewage Sludge*, DoE, London, 1989.

Table 2 Destination of sludge in tonnes dry weight

Destination	1992	1998 (Forecast)	2005 (Forecast)
Disposal to surface waters by pipeline	9609	3211	0
Disposal to surface waters by ship	273 158	195 900	0
Recycling to agriculture	467 824	604 930	913 870
Re-use (forestry, land reclamation)	35 110	52 770	75 630
Landfill	129 754	75 972	95 311
Incineration	89 800	138 600	415 700
Other (ill defined)	24 300	2690	2760
Total	1 029 555	1 074 073	1 503 271

Disposal of Sludge. The options for disposal of sewage sludge are limited to:

- Recycling to agricultural land
- Other beneficial uses in forestry, land reclamation, compost production
- Landfill
- Thermal destruction

Disposal by dumping at sea came to an end in Europe on 31st December 1998 as a requirement of the Urban Waste Water Treatment Directive (UWWTD). In 1998, about 50% of the UK's sewage sludge was recycled to land, 15% taken to landfill, 9% incinerated and 17% dumped to sea.[9] The volumes and destinations of sludge are expected to change considerable by 2005, as shown in Table 2.[7]

These predicted changes in volumes and destinations are the result of a number of drivers, including:

- The end of dumping to sea on 31st December 1998
- The increase in the proportion of the population that receives waste water treatment as a result of the UWWTD
- The decreasing acceptability and increasing costs of landfill
- The increased availability of high-technology solutions such as incineration

What are Biosolids and what is Recycling? The UK Government[10] considers that recovering value from sewage sludge through spreading on agricultural land is the best practicable environmental option for sludge in most circumstances. A United States Environmental Protection Agency (EPA) definition of biosolids is 'primarily treated waste water materials from municipal waste water treatment plants that are organic in nature and suitable for recycling as a soil amendment'. Sewage sludge refers to untreated primary and secondary organic solids from the wastewater treatment process.[11]

[9] Environment, Transport and Regional Affairs Committee, *Second Report, Sewage Treatment and Disposal*, The Stationery Office, London, 1998, p. 35.

[10] Department of the Environment, Transport and the Regions, *Raising the Quality*, paragraph 71, DETR, London, 1998.

[11] US EPA, *Water Biosolids/Sewage Sludge Use and Disposal*, 2000, http://www.epa.gov/owm/genqa.htm, p. 1.

The benefits of applying biosolids to agricultural land include:[11]

- Slow release of nitrogen
- Release of phosphorus, potassium and essential micro-nutrients, including zinc and iron
- Liming properties, if treated with lime
- Improvement of water-holding capacity, soil structure and air and water transport

Despite these identified benefits, the biosolid recycling route is under threat for a number of reasons:

- Concern raised by a number of stakeholders, including the general public, retailers and users of biosolids, regarding the risk of pathogenic diseases being transmitted to crops, animals and man
- The risk of contaminating land with potentially toxic elements (PTEs) and organic contaminants and associated liabilities for that contamination
- The consequences of transmitting emerging and unknown hazards including, for example, dioxins for all stakeholders in the recycling process
- Public perception that recycling is an inherently unacceptable practice
- Possible nuisance associated with transport of biosolids, including additional vehicle movements, vehicle emissions and odour during transport
- Possible nuisance during biosolid application, including odour and vehicle movements
- Possible reduction of soil micro-organism biodiversity
- Risk of surface and groundwater pollution
- Farmer acceptability on grounds of cost, ease of application and limitations placed on land use, by comparison with other soil conditioners and sources of nutrients
- Increased pressure within supplier organizations to recover costs for biosolid treatment

3 The Legislative Framework and Industry Practice in the UK and USA

UK

The most suitable measures for controlling transmissible diseases during biosolid recycling in the UK are:[12]

- Physical barriers such as injecting or burying sludge
- Treatment or storage, including disinfection, to reduce pathogen numbers and infectivity
- Timing restrictions on land use to allow further reductions in pathogen numbers and infectivity
- Reduction in attractiveness of sludge to vectors, including rodents and insects, through stabilization

---

[12] Ref. 4, p. 41.

- Restrictions for the use of sludge on crop types, in particular those that are eaten raw or are in direct contact with sludge

In the UK the use of sewage sludge on agricultural land is controlled through Regulations (SI 1989, No 1263), accompanied by the non-statutory Code of Practice,[8] where the Regulations implement EC Directive 86/278/EEC on the protection of the environment, and in particular of the soil, when sewage sludge is used in agriculture. The objectives of the Directive (86/278/EEC) include the prevention of harmful effects on soil, vegetation, animals and man, thereby encouraging the correct use of such sewage sludge. The Directive requires that:

- Limits are set for heavy metals
- Sludge is treated before use (where treated sludge has undergone biological, chemical or heat treatment, long-term storage or any other appropriate process so as significantly to reduce its fermentability and the health hazards resulting from its use), although raw sludge can be injected or worked into the soil
- Comprehensive and up-to-date records are kept
- There is no grazing or harvesting of forage crops within 3 weeks of sludge use
- There is no harvesting within 10 months of fruit or vegetables which are normally in direct contact with the soil and normally eaten raw
- No sludge is used on soil in which fruit or vegetable crops (other than fruit trees) are growing

Sludge and soil testing is required by the Directive for pH, various heavy metals, dry matter, organic matter, nitrogen and phosphorus. The Regulations specify the frequency of testing. The Directive also requires that nutrient needs of plants and water quality be taken account of. The Code of Practice is intended to ensure that sludge is compatible with good agricultural practice and that human, animal and plant health are not put at risk.

The approach employed in the UK to implement the Directive through the Code of Practice is to impose barriers to the recycling of pathogens from sludge to their hosts. The approach employed is to:

- Either significantly reduce pathogen numbers by at least 90%, followed by restrictions on grazing, cropping or harvesting to allow an additional reduction in pathogen numbers
- Or avoid direct contact with pathogens by injection of sludge into grassland or cultivation of sludge into arable land followed by restrictions on land use to allow an additional reduction in pathogens

The efficacy of this approach depends on the pathogen:

- Rate of shedding from the host
- Infection rate
- Survivability during sludge treatment
- Survivability in soil

In a review of the Code of Practice it was found that the barrier approach was sound and will reduce risks to an acceptable level for all foreseen eventualities at the time of its formulation in 1989.[13] However, the approach employed was developed as a result of extensive reviews, at the time, of the control of *Salmonella* spp., *Taenia saginata* (beef tapeworm) eggs and *Globodera* spp. (potato cysts nematode). Additional subsequent research has shown that *Taenia saginata* has a higher survivability than 10% when subject to mesophilic anaerobic digestion at temperatures of 25 °C, as allowed by the Code of Practice. Since 1989 a number of pathogens have been recognized as potential sources of illness and include enterotoxigenic *Escherichia coli*, rotaviruses, small round viruses, *Cryptosporidium* spp. and a number of other bacteria. Therefore there is some scientific concern over the validity of the protection afforded by the Code of Practice.

In order to reassure the UK food industry and the British Retail Consortium in particular, a Safe Sludge Matrix was developed in 1998[14] that helped address continuing concerns, including:[15]

- The potential for pathogen transfer in sewage sludge used on agricultural land
- Public concerns, and those of retailers, about human and animal health

The Safe Sludge Matrix (Table 3) greatly expands the crop limitations defined in the Code of Practice and addresses survivability during treatment and subsequent use of sludge. The Matrix:

- Bans use of all untreated sludge (effective since 31st December 1999), except for certain combinable crops subject to further heat processing
- Bans surface application of treated sludge on grassland (effective since 31st December 1999)
- Defines two levels of treatment according to use

The legislation controlling agricultural use of biosolids is presently undergoing further review in the UK. New Regulations are being developed that take account of the Safe Sludge Matrix, revisions to the Sludge Directive (86/78/EEC) and the inclusion of the principle of CCPs derived from HACCP.

USA

In the USA the approach to the control of pathogens in sewage sludge is broadly similar to the controls employed in the UK and is based on two principles:[16]

- Either reduction in pathogens to below detectable limits through treatment

[13] Ref. 4, p. 79.

[14] ADAS, *The Safe Sludge Matrix, Guidelines for the Application of Sewage Sludge to Agricultural Land*, ADAS, Guildford, 1998.

[15] J. Thatcher, in *Sewage Sludge, Disposal, Treatment and Use*, IBC Global Conferences, 2000, p. 3.

[16] US EPA, *Environmental Regulations and Technology. Control of Pathogens and Vector Attraction in Sewage Sludge*, United States Environmental Protection Agency, Washington, EPA/625/R-92/013, 1999, p. 6.

Table 3 Safe sludge matrix

Crop type	Untreated sludge	Treated sludge (reduction of fermentability and possible health hazard by biological, chemical or heat treatment)	Advanced treated sludge (virtual elimination of pathogens)
Fruit	×	×	✓
Salads	×	× (30 month harvest interval applies)	✓
Vegetables	×	× (12 month harvest interval applies)	✓
Horticulture	×	×	✓
Combinable & animal feed crops	✓ (Target end by 31st December 1999)	✓	✓
Grass—grazing	×	× (Deep injection or ploughed down only)	✓
Grass—silage	×	✓	✓
Maize—silage	×	✓	✓

Table 4 Class A sludge pathogenic criteria

Pathogen	Limit per 4 g dry weight of biosolids[a]
Salmonella spp.	<3 MPN
Enteric viruses	<1 PFU
Viable helminth eggs	<1

[a]MPN, most probable number; PFU, plaque forming units.

- Or prevention of vectors from coming into contact with land-applied biosolids (barrier methods)

The regulation of pathogens in the US EPA's Part 503 rule is not based on a risk assessment as there were, at the time of its formulation, insufficient data to allow a meaningful assessment.[17] Instead, the non-risk based Part 503 pathogen operational standard includes pathogen controls and monitoring requirements. The Part 503 rule allows a combination of monitoring and Processes to Significantly Reduce Pathogens (PSRP) and Processes to Further Reduce Pathogens (PFRP) approaches for controlling pathogen densities in biosolids.[18]

[17] US EPA, *A Guide to the Biosolids Risk Assessments for the EPA Part 503 Rule*, United States Environmental Protection Agency, Washington, EPA832-B-93-005, 1995, p. 50.
[18] Ref. 16, chapters 4, 5 and 6.

Table 5 Six methods of reducing pathogens[a]

Alternative and description	Treatment description	Variations	Vector attraction reduction (see Table 6)
1. Thermally treated sewage sludge	Four different time–temperature regimes in which all the sludge particles meet the required time–temperature targets	More conservative time–temperature regimes are applied to sewage with greater than 7% solids which tend to form internal structures that inhibit mixing	High volatile solids must be addressed by pH treatment or heat drying (Options 6 and 7) or Options 8 to 11
2. High pH–high temperature	pH > 12 for >72 hours; temperature >52 °C for at least 12 hours; and air drying to over 50% dry solids after the 72 hours at pH > 12	Requires batch monitoring of pH and temperature at various points and agitation of sludge	Option 6 is automatically met
3. Other processes	Relies on comprehensive monitoring to show treatment satisfies Class A requirement: Faecal coliforms <1000 MPN/g dry solids; or *Salmonella* spp. <3 MPN/4 g dry solids; and Enteric viruses <1 PFU/4 g dry solids; and Helminth eggs <1/4 g dry solids		Dependent on pathogen reduction process
4. Unknown process (including sludges that have not satisfied the treatment requirements of other Class A alternatives)	At the time of use or disposal the sludge must meet the following: Enteric viruses <1 PFU/4 g dry solids; and Helminth eggs <1/4 g dry solids; and Faecal coliforms <1000 MPN/g dry solids; or *Salmonella* spp. <3 MPN/4 g dry solids		Dependent on pathogen reduction process
5. Use of process to further reduce pathogens (PFRP)—see notes below	Treated according to one of the PFRPs and Faecal coliforms <1000 MPN/g dry solids; or *Salmonella* spp. <3 MPN/4 g dry solids at the time of use or disposal		Dependent on pathogen reduction process

| 6. Use of a process equivalent to a **PFRP**—see notes below | Treated according to a process equivalent to a PFRPs and Faecal coliforms <1000 MPN/g dry solids; or *Salmonella* spp. <3 MPN/4 g dry solids at the time of use or disposal | Dependent on pathogen reduction process |

[a] MPN, most probable number; PFU, plaque forming units.

Processes to Further Reduce Pathogens (PFRPs):

Composting, the controlled aerobic decomposition of organic matter which produces a humus-like material. Sewage sludge is typically mixed with a bulking material and the following operating conditions achieved:

• for the within-vessel or the static aerated pile, the temperature of the sewage sludge is maintained at 55 °C or higher for 3 consecutive days

• for windrow composting, the temperature of the sewage sludge is maintained at 55 °C or higher for 15 consecutive days or longer, during which time the windrow will be turned for a minimum of 5 times

Heat drying where the following operating conditions are achieved:

• sewage sludge is dried by direct or indirect contact with hot gases to reduce the moisture content to 10% or lower. Either the temperature of the sewage sludge particles exceeds 80 °C or the wet bulb temperature of the gas in contact with the sewage sludge as it leaves the dryer exceeds 80 °C

Heat treatment where the following operating conditions are achieved:

• liquid sludge is heated to at least 180 °C (*i.e.* under pressure) for 30 min

Thermophilic anaerobic digestion achieves higher rates of organic solids reduction compared to conventional aerobic processes at ambient temperature; in addition, the digested sludge is effectively pasteurized and is based on the following operating conditions:

• liquid sewage sludge is agitated with air or oxygen to maintain aerobic conditions and the mean cell residence time of the sewage sludge is 10 consecutive days at 55–60 °C

Beta and gamma ray irradiation, under the following operating conditions:

• sewage sludge is irradiated with beta rays from an accelerator at dosages of at least 1.0 megarad at room temperature; and

• sewage sludge is irradiated with gamma rays from certain isotopes, such as cobalt 60 and caesium 137 (at dosages of at least 1.0 megarad) at room temperature

Pasteurization, under the following operating conditions:

• the temperature of the sewage sludge is maintained at 70 °C or higher for 30 min or longer

Table 6 Twelve vector attraction reduction options[a]

Option	Requirement	Typical method
1. Reduction in volatile solids content	Reduction of at least 38% volatile solids during sewage sludge treatment, typically achieved through 'good practice' when anaerobically digested for 15 days at 35 °C	Anaerobic biological treatment and aerobic biological treatment
2. Additional digestion of anaerobically digested sewage sludge	Less than 17% additional volatile solids loss during bench-scale anaerobic batch digestion of the sewage sludge for each additional day at 30–37 °C	Only for anaerobically digested sewage sludge that cannot meet requirements of Option 1
3. Additional digestion of anaerobically digested sewage sludge	Less than 15% additional volatile solids reduction during bench-scale aerobic batch digestion for 30 additional days at 20 °C	Only for aerobically digested liquid sludge with 2% or less solids that cannot meet the requirements of Option 1, e.g. sewage sludges treated in extended aeration plants. Sludges with >2% solids must be diluted
4. Specific Oxygen Uptake Rate (SOUR) for aerobically digested sewage sludge	SOUR at 20 °C $\leqslant$ 1.5 mg O_2/h/g solids dry weight	Liquid sludges from aerobic processes run at between 10 and 30 °C, but not for composted sewage sludges
5. Aerobic processes at greater than 40 °C	Aerobic treatment for at least 14 days at over 40 °C with an average temperature of over 45 °C	Composted sludge
6. Addition of alkali	pH raised by alkali to at least 12 at 25 °C and maintained at pH 12 for 2 h and a pH of 11.5 for 22 h	Sludge treated with alkali including lime, fly ash, kiln dust and wood ash
7. Moisture reduction of sewage sludge containing no unstabilized solids	Contains no unstabilized solids (e.g. food scraps) generated during primary waste water treatment and, if the solids content is at least 75%, by removal of water and not addition of inert materials	Sludges treated by aerobic or anaerobic processes and therefore do not contain unstabilized solids
8. Moisture reduction of sewage sludge containing no unstabilized solids	A dry solids content of 90% will desiccate any unstabilized solids and will limit biological activity. Subsequent wetting may lead to putrefaction and vector attraction and therefore more storage is required.	Sewage sludges containing unstabilized solids generated in primary water treatment, e.g. heat-dried sewage sludge
9. Injection	No significant amount is left on the surface 1 h after injection (if Class A then injection must take place within 8 h after discharge from the pathogen reduction process)	Sludge applied to land or on a surface disposal site including domestic septage applied to agricultural land, forest, reclamation sites or on a surface disposal site

10. Incorporation of sewage sludge into the soil	Sludge incorporated into the soil within 6 h after application to land or placement on surface disposal site; Class A sludge must be applied or placed on the land surface within 8 h after discharge from the pathogen reduction process	Sludge applied to land or on a surface disposal site including domestic septage applied to agricultural land, forest, reclamation sites or on a surface disposal site
11. Covering sewage sludge	Sludge placed on a surface disposal site must be covered with soil or other material at the end of each operating day	Sludge or domestic septage placed on a surface disposal site
12. Raising the pH of domestic septage	Septage pH raised to 12 at 25 °C by alkali addition and maintained at pH 12 for 30 min without further addition of alkali	Domestic septage applied to agricultural land, forest, reclamation sites or on a surface disposal site

[a]Class A sludges are likely to be devoid of actively growing bacteria and are therefore an ideal medium for growth of pathogenic bacteria. If pathogenic bacteria are present, are survivors or are introduced by contamination, their numbers will increase slowly for the first 8 h after treatment. After this period their numbers can increase rapidly. This kind of growth is unlikely to happen for Class B sludges because high densities of non-pathogenic bacteria are present which suppress the growth of pathogenic bacteria

Table 7 Three Class B pathogen requirements	*Alternative and description*	*Treatment description*	*Vector attraction reduction (Options 1 to 12 discussed below)*
	1. Monitoring for faecal coliforms	Geometric mean of 7 samples to be less than 2 million MPN or CFU/g dry solids	
	2. Use of a process to significantly reduce pathogens (PSRP)	See notes below[a]	
	3. Other processes	Relies on comprehensive monitoring to show treatment satisfies Class A requirement: Faecal coliforms $<1000\,MPN/g$ dry solids; or *Salmonella* spp. $<3\,MPN/4\,g$ dry solids; and Enteric viruses $<1\,PFU/4\,g$ dry solids; and Helminth eggs $<1/4\,g$ dry solids	Dependent on pathogen reduction process

[a]Processes to Significantly Reduce Pathogens (PSRPs):
Aerobic digestion of sludge, operated in a variety of ways where the operating conditions are:
  ● sewage sludge is agitated with air or oxygen to maintain aerobic conditions for a specific mean cell residence time at a specific temperature between 40 days at 20 °C and 60 days at 15 °C
Anaerobic digestion where the operating conditions are:
  ● sewage sludge is treated in the absence of air for a specific mean cell residence time at a specified temperature of between 15 days at 35–55 °C and 60 days at 20 °C
Air drying of partially digested sludge where the operating conditions are:
  ● sewage sludge is dried on sand beds or on paved or unpaved basins for a minimum period of 3 months and during 2 of the 3 months the ambient average daily temperature is above 0 °C
Composting where the operating conditions are:
  ● for a within-vessel, static aerated pile, or windrow composting, the temperature of the sewage sludge is raised to 40 °C or higher and remains at 40 °C or higher for 5 days, for 4 h during the 5 day period, the temperature in the compost pile exceeds 55 °C
Lime stabilization where the operating conditions are:
  ● sufficient lime is added to the sewage sludge to raise the pH of the sewage sludge to 12 after 2 h of contact
Equivalent processes are approved by the EPA's Pathogen Equivalency Committee.

For biosolids to be applied to land they must be subject to one of two levels of treatment, referred to as Class A or Class B.

Class A Sludges. Biosolids classified as Class A are subject to treatment designed to reduce pathogens (enteric viruses, pathogenic bacteria and viable helminth eggs) to below the limits of detection shown in Table 4.

  In order to be classified as Class A biosolids and achieve the listed pathogen

Table 8 Site restrictions

Activity/crop	Restrictions
Turf farms	Turf shall not be harvested for 1 year after application if the turf is to be placed on land with high potential for public access
Animal grazing	Animals shall not be allowed to graze on the land for 30 days after application
Public access	Public access to land with high potential for public exposure shall be restricted for 1 year after application
Food crops that do not touch the ground at any time, including harvesting	
Food crops, *e.g.* oats, wheat	Harvest may not take place until 30 days after application
Food crops that do, or may, touch the ground	
Harvested food will not be eaten raw or handled by the public	Permitting authority may use discretion to reduce waiting period from 14 months to 30 days, depending on the application
Harvested food may be eaten raw or handled by the public	
Food crops with harvested parts that touch the sewage sludge/soil mixture but are totally above the land surface, *e.g.* lettuce, cabbage, melons, strawberries and herbs	Harvest may not take place until 14 months after application
Food crops with harvested parts below the land surface	Harvest may take place after 20 months after application when the sludge remains on the surface for 4 months or longer before incorporation into the soil
Food crops with harvested parts below the surface and the sludge is incorporated in less than 4 months after application	Harvest may not take place until 38 months after application

reduction, they must also have been subject to one of six prescribed methods of reducing pathogens as shown in Table 5. The six pathogen reduction alternatives shown in Table 5 are often part of a process that includes vector attraction reduction options. The EPA identifies a total of 12 vector attraction reduction options as shown in Table 6.

Class B Sludges. Class B sludges contain pathogens at levels that are unlikely to pose a threat to public health and the environment under specific use conditions. In addition, they must meet one of the vector attraction reduction requirements (see Table 7). Class B biosolids have reduced densities of pathogenic bacteria and

Table 9 Class A and Class
B application controls

Application	Pathogen requirements	Vector attraction reduction requirements
Bulk application to agricultural land, public access site, including parks and reclamation sites, including mine spoil	Class A or Class B with site restrictions	Options 1–10
Bulk application to a lawn or home garden	Class A	Options 1–8
Biosolids sold or given away in a bag or container for application to land	Class A (Class B cannot be used)	Option 1–8

enteric viruses as demonstrated by a faecal coliform density of 2 million MPN or CFU/g dry solids. Viable helminth eggs are not necessarily reduced in Class B biosolids. There are three Class B pathogen requirements as shown in Table 7.

Site use restrictions are imposed to minimize the potential for human or animal exposure to Class B biosolids for a period of time following land application and until environmental factors, including sunlight and desiccation, have further reduced pathogens (see Table 8). These site restrictions are applied in order to allow time for further reduction in the pathogen population and are primarily based on the survival rates of helminth eggs.

Application Controls for both Class A and Class B Biosolids. Table 9 summarizes the options available for the application of Class A and Class B biosolids.

4 Principal Pathogens in Sewage Sludge

Recognized Pathogens in Sewage Sludge

Human excreta-related diseases are primarily due to four groups of pathogens: viruses, bacteria, protozoa and worms (helminths). Of these, the helminths often have a complex multi-host life cycle and where the hosts may be geographically restricted. The UK Code of Practice recognizes that a number of organisms in sewage sludge are of most concern, including salmonellae, beef tapeworm *Taenia saginata*, potato cyst nematodes and a range of viruses.[8] However, there are a large number of pathogens of concern in domestic sewage and sewage sludge as shown in Table 10.

Micro-organisms that may occur in UK sewage sludge, with those of most significant risk in bold, are shown in Table 11.[19]

---

[19] E. G. Carrington, R. D. Davis, J. E. Hall, E. B. Pike, S. R. Smith and R. J. Unwin, *Review of the Scientific Evidence Relating to the Controls on the Agricultural Use of Sewage Sludge. Part 2, Evidence Relevant to Controls on the Agricultural Use of Sewage Sludge*, Report DET 4454/4, Water Research Centre, 1998, p. 23.

Table 10 Principal pathogens of concern in domestic and sewage sludge[20]

Organism	Disease/symptoms
Bacteria	
Salmonella spp.	Salmonellosis (food poisoning)
Shigella spp.	Bacillary dysentery
Yersinia spp.	Acute gastroenteritis (including diarrhoea, abdominal pain)
Vibrio cholerae	Cholera
Campylobacter jejuni	Gastroenteritis
Escherichia coli (pathogenic strains)	Gastroenteritis (pathogenic strains)
Enteric viruses	
Hepatitis A virus	Infectious hepatitis
Norwalk and Norwalk-like viruses	Epidemic gastroenteritis with severe diarrhoea
Rotaviruses	Acute gastroenteritis with severe diarrhoea
Polioviruses	Poliomyelitis
Coxackieviruses	Meningitis, pneumonia, hepatitis, fever, cold-like symptoms, *etc.*
Echoviruses	Meningitis, paralysis, encephalitis, fever, cold-like symptoms, diarrhoea, *etc.*
Reovirus	Respiratory infections, gastroenteritis
Astroviruses	Epidemic gastroenteritis
Calciviruses	Epidemic gastroenteritis
Protozoa	
Cryptosporidium	Gastroenteritis
Entamaeba histolytica	Acute enteritis
Giardia lambia	Giardiasis (including diarrhoea, abdominal cramps, weight loss)
Balantidium coli	Diarrhoea and dysentery
Toxoplasma gondii	Toxoplasmosis
Sarcocystis	
Helminth worms	
Ascaris lumbricoides	Digestive and nutritional disturbances, abdominal pain, vomiting, restlessness
Ascaris suum	May produce symptoms such as coughing, chest pain and fever
Trichuris trichiura	Whipworm, abdominal pain, diarrhoea, anaemia, weight loss
Toxocara canis	Fever, abdominal discomfort, muscle aches, neurological symptoms
Taenia saginata	Beef tapeworm, nervousness, insomnia, anorexia, abdominal pain, digestive disturbances
Taenia solium	Pork tapeworm, nervousness, insomnia, anorexia, abdominal pain, digestive disturbance
Necator americanus	Hookworm disease
Hymenolepsis nana	Taeniasis

[20] Ref. 16, p. 6.

Table 11 Micro-organisms that may occur in UK sewage sludge

Group	Group
Bacteria	Viruses
Salmonella spp.	**Polioviruses**
***Shigella* spp.**	**Coxackieviruses A and B**
***Yersinia* spp.**	**Echoviruses**
Escherichia	**Rotaviruses**
Pseudomonas	Adenoviruses
Clostridium	Reovirus
Bacillus	Astroviruses
Listeria	Calciviruses
Vibrio	Coronavirus
Mycobacterium	Norwalk agent and other small round
Leptospira	structured viruses (SRSV)
Campylobacter	
Staphylococcus	
Streptococcus	
Protozoa	Cestodes
Crytposporidium	***Taenia***
Entamaeba	*Diphyllobothrium*
Giardia	*Echninoccus*
Balantidium	
Toxoplasma	
Sarcocystis	
Nematodes	Yeasts
Ascaris	*Candida*
Toxocara	*Cryptococcus*
Tricuris	*Trichosporon*
Ancylostoma	
Fungi	
Aspergillus	
Phialophora	
Geotrichum	
Trichophyton	
Edidermophyton	

Micro-organism considered by the World Bank to be of concern[21] are presented in Table 12 on p. 62.

New and Emerging Pathogens in Sewage Sludge

Considerable effort has been expended in reviewing the efficacy of utilizing sewage sludge on agricultural land.[4,17,20] Practice in the UK, based on the Code of Practice, is considered to be sound and will reduce risks of infection to an acceptable level in all foreseen eventualities when considering diseases caused by salmonellae, *Taenia saginata* eggs, potato cyst nematodes and a range of

[21] R. G. Feachem, D. J. Bradley, H. Garelic and D. D. Mara, *Sanitation and Disease*, Wiley, New York, 1983, pp. 9–15.

viruses.[22] Nevertheless, owing to findings of recent research, the 1989 Code of Practice may not adequately control both a number of the hazards considered in 1989 and a number of newly identified hazards. These hazards include:

- *E. coli* O157:H7
- Viruses generally
- *Campylobacter*
- *Listeria* spp. and *Listeria monocytogenes* in particular
- *Salmonella typhimurium* phage type DT 104
- *Cryptosporidium*
- *Giardia*
- BSE agent

5   Existing Management Practices in the UK

Record Keeping

In the UK the organizations producing biosolids maintain records as advised in the Code of Practice[8] including:

- The quantities of sludge produced and the quantities supplied for use in agriculture
- Sludge analyses
- Type of sludge treatment
- Soil analysis (for any 15 cm or 7.5 cm deep monitoring samples as well as for statutory samples)
- Estimates of soil metal concentrations prepared for sites where 25 cm (a regulatory requirement) samples have not yet been taken
- The names and addresses of the recipients of the sludge, the location of each site where sludge is applied and the quantity and quality of sludge supplied. These figures should include the quantity and quality of other sludges (if any) used by the farmer but not supplied by the sludge producer concerned. In such cases the sewage sludge producers involved should agree among themselves which of them is to have overall responsibility for monitoring the particular site
- Written evidence given by the relevant government Minister in respect of sludge use on dedicated sites together with any analyses made of soil crops as a result of that advice

Information for each source works should be permanently recorded to monitor the use of sludge in agriculture. A site should be regarded as a field or part of a field to which sludge has been applied.

Monitoring

In order that they can properly utilize sludge it is necessary for farmers to be

---

[22] Ref. 4, p. 79.

Table 12 Pathogenic organisms of concern to the World Bank

Organism	Distribution	Disease/symptoms
Viruses		
Adenoviruses	Worldwide	Numerous
Polioviruses	Worldwide	Poliomyelitis
Echoviruses	Worldwide	Meningitis, paralysis, encephalitis, fever, cold-like symptoms, diarrhoea, etc.
Coxackieviruses	Worldwide	Meningitis, pneumonia, hepatitis, fever, cold-like symptoms, *etc.*
Hepatitis A virus	Worldwide	Infectious hepatitis
Reovirus	Worldwide	Respiratory infections, gastroenteritis
Rotaviruses, Norwalk agent and other viruses	Worldwide	Diarrhoea
Bacteria		
Campylobacter fetus ssp. *Jejuni*	Worldwide	Diarrhoea
Escherichia coli (pathogenic forms)	Worldwide	Diarrhoea
Salmonella typhi	Worldwide	Typhoid fever
Salmonella paratyphi	Worldwide	Paratyphoid fever
Salmonella spp.	Worldwide	Salmonellosis (food poisoning)
Shigella spp.	Worldwide	Bacillary dysentery
Vibrio cholerae	Worldwide	Cholera
Vibrio spp.	Worldwide	Diarrhoea
Yersinia enterocolitica	Worldwide	Diarrhoea and septicaemia
Protozoa		
Balantidium coli	Worldwide	Diarrhoea and dysentery
Entamaeba histolytica	Worldwide	Acute enteritis
Giardia lambia	Worldwide	Giardiasis (including diarrhoea, abdominal cramps, weight loss)

Helminth worms

Ancyclostoma duodenale	Warm wet climates	Hookworm
Ascaris lumbricoides	Worldwide	Digestive and nutritional disturbances, abdominal pain, vomiting, restlessness
Clonorchis sinensis	Southeast Asia	Chinese liver fluke
Diphyllobothrium latum	Mainly temperate	Fish tapeworm
Enterobius vermicularis	Worldwide	Pinworm
Fasciola hepatica	Worldwide in sheep/cattle raising areas	Sheep liver fluke
Fasciolopsis buski	Southeast Asia, mainly China	Giant intestinal fluke
Gastrodiscoides hominis	India, Bangladesh, Vietnam, Philippines	Gastrodiscoidiasis
Heterophyes heterophyes	Middle East, southern Europe, Asia	Heterophyiasis
Hymenolepsis nana	Worldwide	Taeniasis
Metagonimus yokogawai	East Asia, Siberia	Metagonimiasis
Necator americanus	Warm wet climates	Hookworm disease
Opistorchis felineus	USSR, Thailand	Cat liver fluke
Opistorchis viverini		
Paragonmus westermani	Southeast Asia, foci in Africa and South America	Lung fluke
Schistosoma haematobium	Africa, Middle East, India	Schistosomiasis, bilharzia
Schistosoma japonicum	Southeast Asia	Schistosomiasis, bilharzia
Schistosoma mansoni	Africa, Middle East, Central and South America	Schistosomiasis, bilharzia
Strongyloides stercoralis	Warm wet climates	Threadworm
Taenia saginata	Worldwide	Beef tapeworm
Taenia solium	Worldwide	Pork tapeworm, nervousness, insomnia, anorexia, abdominal pain, digestive disturbance
Ascaris suum	Worldwide	May produce symptoms such as coughing, chest pain and fever
Trichuris trichiura	Worldwide	Whipworm, abdominal pain, diarrhoea, anaemia, weight loss

provided with an analysis of the sludge received which should include the following parameters:

- Dry matter %
- Organic matter %
- pH
- Nitrogen total and ammoniacal (% dry solids)
- Phosphorus total (% dry solids)
- The following heavy metals measured in mg/kg dry solids: zinc, copper, nickel, cadmium, lead, mercury, chromium, molybdenum, selenium, arsenic
- Fluoride (mg/kg dry solids)

Results of all soil analysis should also be provided together with the quantity of sludge supplied to each site.

Management Practice Shortfalls

The explicit purpose of record keeping is to maintain sustainability of the recycling process in terms of Potentially Toxic Elements (PTEs) and not for the control of pathogens. Whilst the Code of Practice states that records are necessary and must be maintained up-to-date, there is no requirement to use the information gathered in a structured way to manage the recycling process.

6 Hazard Analysis and Critical Control Point

Overview of HACCP

HACCP is a quality assurance system widely used in the food industry to help prevent both food safety and product quality problems. When agreeing to develop an external verified quality assurance system for biosolids recycling, Water UK, the trade association of the UK water industry, did not specify which quality assurance system would be appropriate. However, given that HACCP is the preferred management system of the UK food industry and the key stakeholder concern is food safety, HACCP is the natural choice.

Product Safety and Product Quality

Historically, sludge has been regarded as a waste to be disposed of, but it is increasingly regarded as a resource of positive value. Indeed the Royal Commission considers the process should be reviewed with the aim of enhancing its quality and safety.[23] To date the management of the biosolid process has been concerned with issues of product safety and the presence of pathogens and PTEs in particular. The application of HACCP to biosolid recycling will provide an opportunity to address the control of both safety and quality.

[23] Ref. 7, p. 95.

Table 13 Seven principles of HACCP

Principle	Activity
1	Conduct a hazard analysis
2	Determine the critical control points (CCPs)
3	Establish critical limit(s)
4	Establish a system to monitor control of the CCP
5	Establish the corrective action to be undertaken when monitoring indicates that a particular CCP is not under control
6	Establish procedures for verification to confirm that HACCP is working effectively
7	Establish documentation concerning all procedures and records appropriate to these principles and their application

The Seven Principles of HACCP

The seven principles of HACCP,[24] shown in Table 13, are designed to:

- Identify specific hazards that adversely affect the safety of food
- Specify measures for their control

These seven principles will be addressed in the development of a HACCP plan through 14 stages as shown in Table 14 and described in the following sections.[25] The generalized requirements for a biosolids HACCP plan are shown in Table 14 and in the hypothetical HACCP plan for biosolid cake shown in Table 15.

The Fourteen Stages of HACCP

Terms of Reference (Stage 1). In order to limit the extent of the HACCP plan, the terms of reference for the HACCP study should be clearly defined:

- The limit of the process to be subjected to the HACCP process
- The geographical location of the study, including details of the sites included in the study
- The intended use of the product and therefore the categories of hazard to be addressed including biological, chemical and physical hazards
- The point at which the product is judged to be safe with accompanying storage and usage instructions

Selection of the HACCP Team (Stage 2). A multi-disciplinary team should be identified to develop, maintain and review the HACCP plan.

Product Description (Stage 3). In order to address both product safety and quality the starting point of a HACCP plan is to define the product. Safety issues

[24] Codex Alimentarius Commission, *Food Hygiene Basic Texts*, Food and Agriculture Organisation of the United Nations, Rome, 1997, pp. 35–36.
[25] S. Leaper, ed., *HACCP: A Practical Guide*, 2nd edn., Campden and Chorleywood Food Research Association, 1997, p. 8.

Table 14 Fourteen stages of developing a HACCP plan for biosolid cake

Stage	Activity (corresponding Codex Principle)	Biosolid HACCP activity
1	Define terms of reference	• What are the biosolids to be used for? • Where are the biosolids produced and recycled? • What types of hazards are to be considered (biological, chemical, physical)? • How will the biosolids be judged safe for use? • Define the point when the biosolids are considered safe when accompanied by appropriate storage and use instructions for the farmer
2	Select HACCP team	The HACCP team will be multi-disciplinary and with skilled staff representing at least: • Trade effluent control • Sewage treatment works operation • Advanced treatment operation • Biosolids transportation • Biosolids recycling • Quality assurance • Regional supervisors • Environment, Health and Safety specialists • Representation of sub-contractors • HACCP specialist (chairperson)
3	Describe the product	The description will address: • The fact that the raw material is waste water and possibly includes trade effluents and imported sludges • Describe the broad sludge treatment process employed • Any packaging employed for the biosolids • Any storage and distribution conditions required for the biosolids • The maximum period before which the biosolids can be used for their intended purpose • Any instructions for use of the biosolids including the requirement to abide by the Safe Sludge Matrix • The nature of the final product in terms of pathogen, PTE, dry solid and litter content and the potential of the biosolids for odour
4	Identify intended use	A description of the use of biosolids as a fertilizer and soil conditioner
5	Construct a flow diagram	The process diagram will typically compress all of the process steps An example process flow diagram is presented at Table 15 (this diagram includes additional compression to avoid repetition)
6	Confirm flow diagram	A key step in order to ensure the flow diagram contains no errors or omissions
7	List potential hazards and identify control measures (1. Conduct a hazard analysis)	For each process step, identify potential hazards. These hazards relate to safety (regulatory) or quality (operational) hazards and can be divided into microbiological (*e.g.* pathogenic bacteria, viruses, protozoa and helminth worms), chemical (PTEs) or physical (dry solids, litter and odour potential)

Biosolid Recycling and Food Safety Issues

8

Determine CCPs
(2. Determine the critical control points)

Apply the decision tree at Figure 1 (see below) to determine if the identified hazard is a CCP
The regulatory CCPs, those relating to pathogens and PTEs, are relatively few in number.
Quality CCPs, typically relating to physical hazards, are likely to be more numerous.
Quality CCPs often relate to ensuring a sufficiently high dry solid content during the several steps of the treatment process and the avoidance of odour and litter complaints by the end user

9

Establish critical limits for CCPs
(3. Establish critical limits(s))

Derive critical limits from at least:
- Code of Practice (and new legislation when available)
- Safe Sludge Matrix
- EPA Part 503 Rule
- Research programmes
- Consultation papers, draft regulations, draft Directives
- Information arising from the results of expert workshops
- Any other source agreed as appropriate by the HACCP team and satisfying the criteria of best industry practice

Some typical limits are:
- For digestion temperature: $35 \pm 3\,°C$
- For digestion residence: mean minimum of 12 days
- For trade effluents: PTE concentrations as per trade effluent consents
- For thickening: dry solid content sufficient in order not to threaten the efficient running of the subsequent stages of the process
- For screening: no complaints from farmers

10

Establish monitoring system for each CCP
(4. Establish a system to monitor control of the CCP)

Define a system to monitor the control of the CCP wherever possible using parameters other than microbiological
Typical monitoring parameters are time, temperature, and visual inspection:
- For digestion temperature in each digester, time of batch feeding, mixing gas feed
- For trade effluents, monitoring PTEs as part of consent process
- For thickening, visual inspection and/or dry solids analysis
- For screening, visual inspection

Define frequency of monitoring recognizing that it is intended to show the control is in place and within specification. Define who will undertake the monitoring

11

Establish corrective action plan for each CCP
(5. Establish the corrective action to be undertaken when monitoring indicates that a particular CCP is not under control)

Define the actions to be taken when a CCP is outside, or trending outside, its critical limit
These typically include activities such as:
- Review of the process
- Review of what will be done with biosolids that are produced when the CCP is out of control

67

Table 14 (*cont.*)

Stage	Activity (*corresponding Codex Principle*)	Biosolid HACCP activity
12	Verification (6. Establish procedures for verification to confirm that HACCP is working effectively)	Verification differs from monitoring, being required to show that the entire HACCP plan is working correctly. Verification activities include: • Internal and external audits and the retention of the records • End-product testing to show that the sludge treatment process is effective • Short-term testing to demonstrate that monitoring of surrogate parameters is adequate • Review of deviations from critical limits • Review of corrective actions undertaken • Review of the use of biosolids and associated safety and quality related issues in the wider market place • Audit of records and associated procedures to monitor if CCPs are under control
13	Establish documentation and record keeping (7. Establish documentation concerning all procedures and records appropriate to these principles and their application)	Documentation will include: • HACCP plan • Relevant procedures, including any specially prepared for the HACCP plan • Existing procedures in Operations and Maintenance Manuals, Work Instructions, Total Quality Management systems, quality management systems and any other source of relevant procedures Records will include: • Records of inputs including waste water, trade effluents, sludges, cess pit wastes and treatment chemicals • Records of deviations, corrective actions, modifications to the process • Verification data • Monitoring data The length of time which the various records will be kept for should be determined and should reflect the hazards addressed
14	Review HACCP plan	A periodic review of the HACCP plan should be undertaken on a regular basis, typically annually. Any changes to at least the following list will require an appraisal of the requirement for a review of the HACCP plan • Trade effluent regime • Imported sludge regime • Waste water catchment • Use of treatment chemicals • Process steps or equipment within a process step • Storage • Packaging • Staffing levels • Farmer use • Legislation • Market intelligence regarding safety or quality issues • Emergence of new hazards

are clearly defined in the relevant legislation concerned with biosolids and relate to biological (pathogens) and chemical (PTEs) hazards. Quality issues are specific to each biosolid producer and typically relate to physical hazards (dry solid content, presence of litter and odour potential). Therefore the key parameters in any product description are likely to include:

- Pathogen content
- PTE content
- Dry solids content
- Presence of litter
- Potential for odour generation

The HACCP process also includes a number of other parameters that address, for example:

- The sludge treatment process employed
- Storage duration and conditions
- Any packaging employed
- Any specific instructions for use and application

These parameters have a direct bearing on the safety and quality of the product.

Intended Use (*Stage* 4).   Having described the product, the intended use of the product must also be defined, typically as a fertilizer and soil conditioner.

The Process Flow Diagram (*Stage* 5).   A process flow diagram is a simple way to facilitate hazard analysis during the HACCP process and should include all those steps at which hazards are identified at Stage 7. Initially, process steps are included along with any addition of treatment chemicals or packaging. Subsequently, these process steps may consolidated if considered appropriate. The process flow diagram at Table 15 presents such a consolidated flow diagram.

Confirmation of Process Flow Diagram (*Stage* 6).   The process flow diagram may contain errors, overlooked short circuits in the process flow, missing steps, overlooked raw material inputs or outputs. It is important that the flow diagram is subject to rigorous scrutiny and agreed as being a correct and a valid representation of the process.

Identification of all Potential Hazards and Identification of Control Measures (*Stage* 7).   A brainstorming session is typically undertaken to identify all possible hazards to the safety and quality of the final product. These hazards include those present in the raw materials, hazards that may be introduced during the process and hazards that survive each process step. Generally these hazards can be categorized as microbiological, chemical or physical.

From the forgoing discussion an extensive list of hazards might be recognized:

- Microbiological: all those pathogens identified in Section 4

Table 15 Simplified hypothetical HACCP plan for biosolid cake[a]

Process step	Hazard	Control	CCP? (logic tree)	Critical limits	Monitoring	Corrective actions	Records
1. Trade effluent discharge	Presence of PTEs	Trade effluent consent	Yes (Regulatory) (Q2 'Yes')	Compliance with consent	According to consent conditions, or annually	Review trade effluent consent Review product compliance	Consent monitoring CAs undertaken
2. Waste water arrival	Presence of microbiological hazards	Digestion at step 6	No (Q4 'Yes')				
	Presence of litter	Screening at step 3	No (Q4 'Yes')				
	Presence of PTEs	Field application at step 10	No (Q4 'Yes')				
3. Screens	Litter passing screens	Screening of effluent	Yes (Quality) (Q2 'Yes')	Sufficient to avoid farmer complaint	Daily VI of screens	Repair, unblock, redesign screens Review product acceptability to farmers and determine action for out-of-specification product	VI records Farmer complaint records CAs undertaken
4. Primary and secondary waste water treatment	Survival of microbiological hazards as a result of inadequate digestion	Low dry solids due to incorrect dosing and treatment at several stages	Yes (Quality) (Q4 'No')	No compromise of later stages of the process	Daily VI of settlement and dosing Daily dry solid analysis	Adjust process as necessary Review product acceptability to farmers and determine action for out-of-specification product	VI and dry solid records CAs undertaken
5. Polyelectrolyte dosing of settled sludge							

Process step	Hazard	CCP	Critical control question	Critical limit	Monitoring	Corrective action	Records
6. Thickening	Survival of microbiological hazards as a result of inadequate digestion	Correct dosing at step 5	Yes (Quality) (Q4 'No')	Sufficient dry solids to ensure effective digestion	Daily VI Daily dry solid analysis	Adjust process as necessary Review product acceptability to farmers and determine action for out-of-specification product	VI and dry solids records CAs undertaken
7. Digestion	Survival of microbiological hazards	Correct temperature	Yes (Regulatory) (Q2 'Yes')	35 ± 3 °C	Temperature monitoring every shift	Adjust process as necessary Review product acceptability to farmers and determine action for out-of-specification product	Temperature records CAs undertaken
		Efective mixing		Sufficient to ensure effective digestion	Gas flow to mixers every shift	Adjust process as necessary Review product acceptability to farmers and determine action for out-of-specification product	Gas flow records CAs undertaken
		Correct duration[b]		Mean minimum 12 days	Monitoring batch feed every shift	Adjust process as necessary Review product acceptability to farmers and determine action for out-of-specification product	Batch feed records Corrective actions undertaken
8. Polyelectrolyte dosing							
9. Thickening	Incorrect dry solid content	Correct dosing at step 8	Yes (Quality) (Q4 'No')	Dry solid content as defined in product description	VI/dry solid analysis once per day	Adjust process as necessary Review product acceptability to farmers and determine action for out-of-specification product	VI and dry solid analysis records CAs undertaken

Table 15 (cont.)

Process step	Hazard	Control	CCP? (logic tree)	Critical limits	Monitoring	Corrective actions	Records
10. Cake storage	Survival of biological hazard	Correct storage time	Yes (Quality) (Q2 'Yes')	14 days storage	VI every 14 days	Review procedure Review product acceptability to farmers and determine action for out-of-specification product	VI CAs undertaken
11. Field application	Exceed PTE limits	Correct field application rates	Yes (Regulatory) (Q2 'Yes')	PTE limits as defined in the Code of Practice	Monthly audit PTE monitoring	Review procedures Review product acceptability	PTE monitoring records CAs undertaken
	Survival of biological hazard	Farmer to agree to abide by Safe Sludge Matrix	Yes (Regulatory) (Q2 'Yes')	Application to approved crops	VI on a cycle determined by risk assessment	Withdraw supply of biosolids	Evidence that farmer has agreed to Safe Sludge Matrix VI records CAs undertaken
	Presence of litter	Screening at step 3	No (Q3 'No')				

aVI, visual inspection; CAs, corrective actions.
bEffective volume of the digester will be determined as a result of a 1 in 10 year lithium test, effective mixing volume will be inferred from effective volume.

- Chemical: PTEs, substances introduced during the waste water treatment process, the 200 pollutants considered by the EPA[26]
- Physical: dry solid content, presence of litter, potential for odour nuisance.

At this stage of the HACCP process, no attempt is made to define which of these hazards will generate Critical Control Points in the HACCP plan. Having identified all potential hazards, an analysis is undertaken to determine which should be eliminated or reduced to acceptable levels in order to ensure the product is safe and of a suitable quality. This process will consider:

- Past history of the hazard resulting in safety and quality related effects
- The severity of the consequences of the hazard including short- or long-term, acute or chronic, lethal or sub-lethal
- The consequence of the hazard and effects on product quality
- The size of the population exposed to the hazard
- Vulnerability of those exposed
- Survival and potential for re-growth of micro-organisms of concern
- Potential for secondary hazards to be produced, including generation of toxins, chemical or physical hazards

This step relies on the understanding and professional capabilities of the HACCP team. It is therefore crucial that the HACCP team includes capable individuals familiar with best working practice in the industry and are involved in the whole of the sewage sludge and recycling process.

Having identified the hazards the HACCP team will then determine what control measures can be applied to each hazard. There may be hazards for which there are no control measures in existence. Control measures are intended to prevent, eliminate or reduce hazards to an acceptable level.

Identification of Critical Control Points (CCPs) (Stage 8). In order to make the selection of CCP non-subjective, the Codex Alimentarius Commission of the Food and Agriculture Organization[27] has designed a decision tree to aid in their identification (Figure 1). This decision tree is applied at each step of the process where a hazard has been identified.

Critical Limits for each CCP (Stage 9). Critical limits are required for the identified CCPs. These critical limits can be derived from a number of sources, including legislation and from experimental data. For the biosolid and recycling process, possible sources of critical limits could be:

- Code of Practice
- Safe Sludge Matrix
- EPA Part 503 Rule
- Research programmes
- Consultation papers, draft regulations, draft Directives

[26] Ref. 17, p. 10.
[27] Ref. 24, p. 44.

Figure 1 CCP decision tree

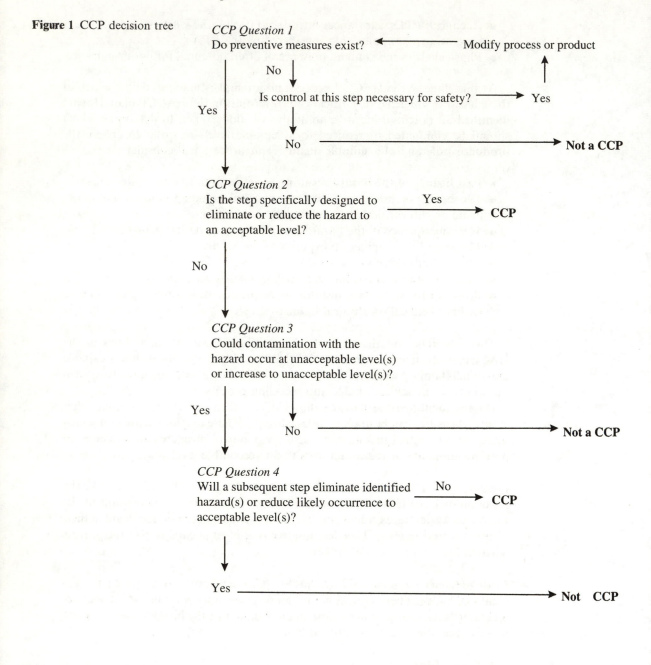

CCP Question 1

Do preventive measures exist? ←—————— Modify process or product

No ↓

Is control at this step necessary for safety? ——————→ Yes

Yes

No ——————————————————————————————→ **Not a CCP**

CCP Question 2

Is the step specifically designed to   Yes
eliminate or reduce the hazard to ——————→ **CCP**
an acceptable level?

No

CCP Question 3

Could contamination with the
hazard occur at unacceptable level(s)
or increase to unacceptable level(s)?

Yes

No ——————————————————————————————→ **Not a CCP**

CCP Question 4

Will a subsequent step eliminate identified   No
hazard(s) or reduce likely occurrence to ——————→ **CCP**
acceptable level(s)?

Yes ——————————————————————————————→ **Not   CCP**

- The results of expert workshops
- Any other source agreed as appropriate by the HACCP team and satisfying the criteria of best industry practice

Critical limits may have a tolerance, such as dry solids content for biosolid cake of 21% with a target of $23 \pm 2\%$, or temperature during digestion of 32 °C with a target of 35 ± 3 °C.

In addition, critical limits must be easily measurable and in the case of the biosolids process these typically include parameters such as temperature, time, dry solid content, PTE content and parameters measured by visual inspection.

Monitoring System for Each CCP (Stage 10). Having selected critical limits for each CCP, wherever practicable that are easily measured, a monitoring system should be defined that monitors control of that CCP and addresses the following requirements:

- Data are generated in a form suitable for subsequent auditing
- Provision of information on a time-scale that enables timely action to bring a CCP back under control (microbiological monitoring rarely fulfils this criterion and surrogates should be employed wherever practicable such as temperature and duration of treatment process)
- Match frequency of monitoring with duration of treatment step in order to ensure bulk products such as biosolids are monitored appropriately (daily monitoring of sludge digestion duration but continuous monitoring of a sludge dryer feed rate)
- Ensure that responsibility for undertaking monitoring is defined and that data are quality assured
- Monitoring requirements are supported by appropriate documented procedures

Corrective Action Plans (Stage 11). Corrective actions are to be identified for each CCP and undertaken when monitoring has shown that the CCP is out of control or showing a trend toward loss of control. Typical corrective action plans include:

- Adjusting equipment
- Reviewing procedures
- Action to be taken when biosolids are produced whilst the CCP was out of control

Verification (Stage 12). Verification differs from monitoring. While monitoring is concerned with the critical limits associated with individual CCPs, verification is concerned with the correct functioning of the whole process and includes:

- Internal and external auditing of the HACCP plan and its records
- Microbiological examination of intermediate and final product
- Validation of the use of surrogate monitoring parameters

- Review of health related issues associated with the treatment and use in the wider market place
- Review of consumer use of the product
- Audit of records and associated procedures to monitor if CCPs are under control

Documentation and Record Keeping (*Stage* 13).   As for any management system, the HACCP plan requires appropriate documentation and record keeping in order to be able to demonstrate that it satisfies the criteria of the Codex Alimentarius and that the plan is up to date and being applied effectively. This will include the following documents:

- A HACCP manual, that can be made part of an existing management system if required
- Existing procedures from a number of sources including, for example, Operation and Maintenance Manuals, Work Instructions and regulatory reporting requirements

Additional record keeping will typically include:

- Source and condition of raw materials
- Process records
- Management decisions relating to HACCP and product safety
- Process deviations
- Corrective actions
- Process modifications
- Verification and validation data
- Review data

The timescale for the maintenance of records should be defined in the HACCP plan.

Review of the HACCP Plan (*Stage* 14).   Periodic review should be undertaken in proportion to the risk associated with the use of the final product. A review of the HACCP plan should be undertaken if there are changes in at least the following issues:

- Raw material make-up
- Equipment making up the process system or modification to that equipment
- Process flow diagram
- Suppliers or supplied raw materials
- Storage arrangements including duration and packaging
- Staffing levels or responsibilities
- Use of product including new outlets
- Understanding of the risks associated with the use of the product
- New hazards
- Legislation

The introduction of new technology will require the HACCP plan to be reviewed.

The Benefits of Applying HACCP to the Recycling Process

Experience of the application of HACCP to the recycling process in the UK has shown that it offers a range of benefits, a number of which are common to any process. The following benefits are particularly relevant to the biosolid recycling process:

- Addresses food safety issues
- Addresses product quality issues
- Encourages the view that biosolids have economic value rather than being a waste requiring potentially expensive disposal
- Focuses attention at the critical parts of the process
- Reducees the quantities of out-of-specification product produced through the application of a preventative approach to process control
- Provides useful support for a defence of 'due diligence' in the event of a food related health incident
- Provides reassurance to stakeholders (including food consumers, water company customers, the food industry and sludge producer's colleagues and directors) that biosolids are being properly managed
- Identifies all currently conceivable hazards, including those which can reasonably be predicted to occur

7 Conclusions

Sewage sludge accounts for approximately 1% of the organic wastes spread to agricultural land, with the remainder accounted for by agricultural and exempt wastes, both of which are unregulated at the time of writing.

The volumes of sewage sludge are increasing dramatically as a result of the implementation of European Directives and legislation, including the Urban Waste Treatment Directive which banned the dumping of sludge to sea on 31st December 1998. In the UK, projections indicate that the quantities will have increased by approximately 50% between 1992 and 2005. This projected increase places the existing outlets, including recycling to agricultural land, under considerable pressure. Whilst recycling is recognized by a number of organization and authorities as being beneficial, there are a number of concerns regarding the presence of pathogens in sewage sludge.

In the USA and UK there are complex systems for the regulation of recycling based on the significant reduction of pathogens in biosolids and ensuring there are physical barriers to avoid direct contact with pathogens. The formulation of these systems has been based on a degree of risk assessment, although in the USA the Environment Protection Agency has approached the management of biosolid recycling on a non-risk-based approach as there are insufficient data to allow a meaningful assessment. In the UK the system was developed based on available information at the time of its formulation in 1989. Review of the available scientific data suggests that the UK system of control is broadly effective but that there are new and emerging pathogens that may not be adequately controlled.

Whilst the quantity of biosolids to be recycled is increasing, the control system has been shown not to be completely reliable. In response to this pressure and concerns conveyed by such organizations as the British Retail Consortium with respect to food safety, the UK Water Industry has undertaken a number of activities. These activities include the development of the Safe Sludge Matrix, a control on recycling more stringent than that required in legislation. In addition, the water industry in the UK as a whole has given a commitment to develop a quality assurance system, that is externally verified, for the control of the biosolid recycling process.

A management system for the biosolid recycling process should address a number of issues including food safety, regulatory requirements and the quality of the biosolids produced. Existing management in the UK is focused on controlling PTEs. Any management system employed should provide the opportunity to effectively manage the process and control food safety issues, regulatory requirements and the quality of the biosolids produced.

HACCP is a management system widely used in the food industry. Its application to biosolid recycling has a number of benefits including a move away from a reactive, end-product testing-led approach to a system of pro-active process optimization. This approach mirrors a change in attitude from treating sludge as a waste by-product requiring expensive disposal to a product of significant value to farmers. As with any product, value can be increased by enhancing quality.

The hazards addressed in a typical biosolids HACCP plan are defined within stage 3 of the HACCP plan and can be divided into:

- Biological hazards relating to pathogens
- Chemical hazards relating to PTEs
- Physical hazards relating to dry solids content, presence of litter and odour potential

These hazards necessarily result in CCPs within the plan which are typically:

- Regulatory CCPs (biological and chemical)
- Quality CCPs (physical)

The identification of quality CCPs allows operational resources to be focused on those critical parts of the sludge treatment process.

In a climate of rapid change the novel application of HACCP to biosolids recycling presents an opportunity to satisfy the requirements of a number of stakeholders:

- Regulators, by providing auditable management of regulatory standards
- The food industry, by addressing food safety issues through the application of a management system widely accepted within the industry
- Farmers, by enhancing the quality of biosolids supplied
- Operational staff, by providing a mechanism for optimizing performance of the process
- Shareholders, by providing evidence of the control of food safety hazards associated with biosolid recycling

Mad Cows, Demented Humans and Food

DAVID M. TAYLOR

1 Introduction

Creutzfeldt–Jakob disease (CJD) and bovine spongiform encephalopathy (BSE) belong to a distinct group of fatal transmissible degenerative encephalopathies (TDEs) of animals and humans (Table 1). The classical, sporadic form of CJD affects around one or two per million of the human population worldwide each year, causing death at a median age of around 66. However, it was reported in 1996 that 10 cases of a new variant form of CJD (vCJD) had been recently identified in relatively young individuals in Great Britain but not apparently elsewhere.[1] This was of widespread interest because (a) the clinical signs of vCJD were unlike those of sporadic CJD, (b) the period of clinical disease before death was much longer than for sporadic CJD, (c) death occurred at a median age of 29, (d) the neurohistopathological features of vCJD were different from those in sporadic CJD, and (e) it concluded that, in the absence of any other obvious cause, the most likely explanation for vCJD was dietary exposure to the agent that had caused a major epidemic of BSE in the UK. This epidemic began around 1985, peaked in 1993, and is currently in steep decline. The incidence of BSE in the UK cattle population has considerably exceeded that in any other BSE-affected country, with more than 177 000 cases having been confirmed by July 2000. By the same time, the number of confirmed vCJD cases in the UK had escalated to 75; three cases had also been identified in France, and one in Ireland. Until 2000, the rate of occurrence of vCJD had been relatively constant, which made it difficult to predict what the eventual scale of this human epidemic might be. However, by July 2000, the number of cases already observed in that year suggested that the incidence was rapidly increasing. This indicated that the eventual scale of the epidemic might be far from modest but that further time would have to elapse to determine whether or not the apparently escalating incidence would continue. Given that this chapter was completed in July 2000, no

---

[1] R. G. Will, J. W. Ironside, M. Zeidler, S. N. Cousens, K. Estibeiro, A. Alperovitch, S. Poser, M. Pocchiari, A. Hofmann, and P. G. Smith, *Lancet*, 1996, **347**, 921.

Issues in Environmental Science and Technology No. 15
Food Safety and Food Quality

D. M. Taylor

Table 1 The transmissible degenerative encephalopathies

Disease	Affected species
Scrapie	Sheep, goats, moufflon
Transmissible mink encephalopathy (TME)	Mink
Chronic wasting disease (CWD)	Elk, mule-deer (in the USA)
Bovine spongiform encephalopathy (BSE)	Cattle, captive exotic ruminants
Feline spongiform encephalopathy (FSE)	Cats, captive exotic felids
Creutzfeldt–Jakob disease (CJD)	Humans
Variant CJD (vCJD)	Humans
Gerstmann–Straussler–Scheinker syndrome (GSS)	Humans
Fatal familial insomnia	Humans
Kuru	Humans

Table 2 The number of cases of indigenous BSE recorded in different countries each year between 1993 and 1999[a]

	1993	1994	1995	1996	1997	1998	1999
Great Britain	34 370	23 945	14 302	8016	4312	3179	2133
Northern Ireland	459	345	173	74	23	18	7
Belgium	0	0	0	0	1	6	3
Denmark[b]	0	0	0	0	0	0	0
France	1	4	3	12	6	18	31
Irish Republic	16	18	15	73	80	83	91
Liechtenstein	0	0	0	0	0	2	0
Luxembourg	0	0	0	0	1	0	0
Netherlands	0	0	0	0	2	2	2
Portugal	3	12	14	29	30	106	170
Switzerland	29	64	68	45	38	14	50

[a]Excluding Canada, Falkland Islands, Germany, Italy and Oman that have reported cases only in imported animals.
[b]One case confirmed in February 2000.

further comments can be made regarding this issue. Studies on the UK cases of vCJD confirmed by July 2000 have, thus far, still failed to implicate anything other than BSE-contaminated foodstuff as the likely cause of their disease. This likelihood is reinforced by the knowledge that TDE agents are able to withstand the temperatures used to cook food by boiling, oven-roasting, pressure-cooking and microwave exposure.[2] The potentially escalating incidence of vCJD has been of particular interest in European countries that appear to have failed, thus far, to have controlled their BSE epidemics (Table 2). It is also the subject of international scientific interest because it represents the first relatively definitive evidence that an animal TDE can transmit to humans. This contrasts with the longstanding experience that occupational or dietary exposure of humans to the agent that causes scrapie in sheep, goats and moufflon does not appear to result in the transmission of scrapie to humans, despite the endemic nature of scrapie in a number of countries.[3]

[2] D. M. Taylor, in *Principles and Practice of Disinfection Preservation and Sterilisation*, ed. A. D. Russell, W. B. Hugo and G. A. J. Ayliffe, Blackwell Scientific, Oxford, 1999, pp. 222–236.
[3] P. Brown, F. Cathala, R. S. Raubertas, D. C. Gajdusek and P. Castaigne, *Neurology*, 1987, **37**, 895.

80

2 The Nature of TDE Agents

TDE agents have been considered to be simply rogue forms of the host's normal prion protein (PrP) that resist catabolic destruction by proteolytic enzymes.[4] This has been considered to result from post-translational structural changes in PrP that are induced by contact with already-modified, disease-specific forms of PrP.[5] The progressive accumulation of the modified, amyloidogenic form of PrP eventually results in the formation of fatal pathological deposits in the central nervous system, within which there is the highest level of expression of PrP. However, the 'protein-only' hypothesis is not universally accepted because it fails to adequately explain the individual phenotypic characteristics of the many strains of TDE agents that can be propagated in mice with the same *PrP* genotype, and the stability of the strain-specific phenotype of the BSE agent in mice regardless of its previous passage history through various mammalian species with different *PrP* genotypes.[6] Some therefore consider that, although it is generally agreed that a modified form of PrP is an essential component of the infectious agents, additional (possibly non-host) informational molecules such as nucleic acids are required to explain strain diversity, but none have yet been identified.[7-10]

3 Routes of Transmission

Genetic

It is considered that some of the TDEs are caused through the inheritance of *PrP* genotypes that results in the spontaneous production of a disease-specific form of PrP, or that spontaneous somatic mutations in the *PrP* gene could have the same effect. The high incidence of scrapie in UK sheep with 'scrapie-susceptible' *PrP* genotypes was considered by some to constitute evidence that it is a genetically transmitted disease, but identical genotypes were subsequently identified in countries where scrapie does not occur.[11] This clearly demonstrated that susceptibility to scrapie, but not the disease itself, is inherited. Also, GSS is a familial human disease that is thought to be inherited as a direct result of the mutation of proline to leucine at codon 102 of the *PrP* gene. However, when transgenic mice were created that exactly reproduced this same single amino acid change, they did not develop a spontaneous neurological disease. They were, however, much more susceptible to experimental challenge with a human-derived GSS agent, compared with other human-derived agents.[12] These types of studies

4 S. B. Prusiner, *Science*, 1982, **216**, 136.
5 G. C.Telling, P. Parchi, S. J. DeArmond, P. Cortelli, P. Montaggna, R. Gabizon, J. Mastrinni, E. Ligaresi, P. Gambetti and S. Prusiner, *Science*, 1996, **274**, 2079.
6 M. E. Bruce, A. Chree, I. McConnell, J. D. Foster, G. Pearson and H. Fraser, *Phil. Trans. R. Soc. London, Ser. B*, 1994, **343**, 405.
7 C. F. Farquhar, R. A. Somerville and M. E. Bruce, *Nature*, 1998, **391**, 345.
8 B. Chesebro, *Science*, 1998, **279**, 42.
9 D. M. Taylor, K. Fernie, P. J. Steele and I. McConnell, *Abstracts of an International Conference on the Characterization and Diagnosis of Prion Diseases in Animals and Man*, Tubingen, 1999.
10 R. A. Somerville, *Trends Microbiol.*, 2000, **8**, 157.
11 N. Hunter, D. Cairns, J. D.Foster, G. Smith, W. Goldmann and K. Donnelly, *Nature*, 1997, **386**, 137.

D. M. Taylor

suggest that TDEs which appear to be inherited may result from the inheritance of disease susceptibility rather than the actual diseases.

Iatrogenic

The accidental transmission of TDEs has occurred through the use of contaminated medicinal products such as vaccines or hormones, or through the use of contaminated surgical instruments or medical devices.[2,13] Such incidents have confirmed what had already been indicated by experimental studies, namely, that disease occurs most rapidly after the introduction of infectivity into the central nervous system. However, peripheral routes (*e.g.* intramuscular or subcutaneous) are also effective but result in longer incubation periods.

Oral/Dietary

With BSE, there is convincing evidence that the epidemic was fuelled by feeding cattle with ruminant-derived meat and bone meal (MBM)[14] that remained infected after processing by the rendering procedures used in the UK during the 1980s.[15] With scrapie in sheep, it is also widely considered that a highly significant factor in the transmission of this disease is the oral exposure of lambs to the infectivity that is known to be potentially present in the placentae or foetal membranes of infected mothers. Exposure to such infectivity is considered to arise post-natally through direct contact or via contaminated bedding, pasture, *etc*. The relevance of this route of infection has been demonstrated by the successful transmission of scrapie to sheep that were dosed orally with foetal membranes from scrapie-affected sheep.[16]

The putative association between cannibalism and the expansion of the kuru epidemic in a tribal population of New Guinea was initially interpreted as definitive evidence that the consumption of kuru-infected brain-tissue was the reason for the expansion of this epidemic. Although there is little doubt that the tissues of kuru victims were removed in some form of reverential ritual, anthropological studies have revealed that visiting research workers never observed the actual consumption of the brains of kuru victims.[17] It was later acknowledged that infection after contact with potentially infected tissues was 'most probably through cuts and abrasions of the skin or from nose-picking, eye rubbing or mucosal injury'.[18]

TME occurs rarely and sporadically, and a foodborne source seems likely but remains unconfirmed. It was claimed that the most recent outbreak (in the USA) occurred in mink that had only been fed on bovine tissues obtained from 'downer'

[12] J.C. Manson, E. Jamieson, H. Baybutt, N.L. Tuzi, R. Barron, I. McConnell, R. Somerville, J. Ironside, R. Will, M.-S. Sy, D.W. Melton, J. Hope and C. Bostock, *EMBO J.*, 1999, **18**, 6855.
[13] P. Brown, M.A. Preece and R.G Will, *Lancet*, 1992, ii, 24.
[14] J.W. Wilesmith, G.A.J. Wells, M.P. Cranwell and J.B.M. Ryan, *Vet. Rec.*, 1988, **123**, 38.
[15] D.M. Taylor, S.L. Woodgate and M.J. Atkinson, *Vet. Rec.*, 1995, **137**, 605.
[16] I.H. Pattison, M.N. Hoare, J.N. Jebbett and W.A. Watson, *Br. Vet. J.*, 1974, **130**, lgv.
[17] W. Arens, *The Man-eating Myth*, Oxford University Press, New York, 1979.
[18] D.C. Gajdusek, *Subviral Pathogens of Plants and Animals: Viroids and Prions*, Academic Press, New York, 1985.

cattle.[19] This was considered to be an indication that cattle in the USA might be infected with a BSE-like disease that was transmissible to mink, but the active cattle surveillance programme in the USA has failed to reveal the presence of any BSE-like disease to date.

Until vCJD emerged, no studies had been conducted to determine the precise mechanisms whereby TDE infections might become established after dietary exposure. In this respect, experimental studies have now been carried out to determine (a) the route(s) by which infection reaches the central nervous system, and (b) which cellular components of the intestinal wall express PrP at sufficiently high levels to represent likely routes of entry for TDE agents. Following the oral challenge of hamsters with hamster-passaged scrapie agent, it has been demonstrated that the earliest appearance of the disease-specific form of PrP in the brain results from transmission through the vagal nerve; infectivity also appears to consistently, but more slowly, reach the brain by traversing through enteric lymphoid tissue, spleen, splanchnic nerves and the thoracic region of the spinal cord, from which it travels to the brain.[20] The latter route has also been found to occur after peripheral infection with mouse-passaged scrapie, and these studies have confirmed the importance of agent replication in the follicular dendritic cells of the spleen before infectivity is delivered to the central nervous system.[21,22] Studies on the distribution of normal PrP in the intestinal wall of humans have demonstrated its presence in nerve endings that form intimate associations with enteric epithelial cells, and show that only a thin layer of epithelium stands between ingested TDE agents and host PrP.[23]

The Occurrence of BSE in Species other than Cattle through Ingestion of the Infectious Agent. Despite the uncertainties regarding the transmission of some other TDEs by a dietary route, there is convincing circumstantial evidence that BSE has transmitted to a variety of species, other than cattle, by this route. In the UK, 87 cases of feline spongiform encephalopathy (FSE) have been identifed in domestic cats since 1990. FSE had not been observed previously and, except for one unexplained case in Norway, has only been observed in the UK. The only apparent route of infection was dietary, and strain typing has shown that the FSE agent is identical to that of BSE.[6] Similarly, BSE has apparently transmitted to a variety of exotic felid and ruminant species housed in, or originating from, UK zoological collections (Table 3). In the case of the felid species, these are likely to have acquired their BSE-like disease through the feeding of putatively BSE-infected bovine tissues. In the case of the ruminant species, their disease probably resulted from feeding them with proprietary feedstuff that contained BSE-contaminated MBM prior to the 1988 ban on such a practice. Confirmatory evidence that these types of diseases were actually caused by the BSE agent has been formally

[19] R. F. Marsh and G. R. Hartsough, *Proceedings of the Seventh Animal Welfare Conference for Feed Animal Veterinary Research*, Arizona, 1986.
[20] M. Beekes, P. A. McBride and E. Baldauf, *J. Gen. Virol.*, 1998, **79**, 601
[21] K. L. Brown, K. Stewart, D. Ritchie, N. A. Mabbott, A. Williams, H. Fraser, W. I. Morrison and M. E. Bruce, *Nat. Med.*, 1999, **5**, 1308.
[22] N. A. Mabbott, F. Mackay, F. Mimms and M. E. Bruce, *Nat. Med.*, 2000, **6**, 719.
[23] A. H. Shmakov, N. F. McLennan, P. A. McBride, C. F. Farquhar, J. Bode, K. A. Rennison and S. Ghosh, *Nat. Med.*, 2000, **6**, 840.

Table 3 Exotic felid and ruminant species housed within, or derived from, British zoological collections that have developed BSE-like diseases

Felid species	Number affected	Ruminant species	Number affected
Cheetah	5	Ankole	2
Ocelot	2	Bison	1
Puma	3	Eland	1
Tiger	1	Gemsbok	1
		Greater kudu[a]	6
		Nyala[a]	1
		Oryx	1

[a]Strain typing has demonstrated that the causal agents are identical to that of BSE.

obtained, at least for kudu and nyala, by demonstrating that the strain of agent causing their diseases was identical to that of the BSE agent.[6] It thus seems beyond reasonable doubt that all of the BSE-like diseases in exotic felid and ruminant species have been caused by the BSE agent through the ingestion of contaminated feed material.

The Association between vCJD and Dietary Exposure to the BSE Agent. Emotive headlines in the UK media during the mid-to-late 1980s suggested that there should be concern about the consumption of beef with regard to the transmission of BSE to humans. However, the product recognized traditionally as beef in the UK would be the muscle tissue of relatively young cattle that were unlikely to have been fed MBM. Even if such cattle were potentially infected, their usual age at slaughter would have meant that they were unlikely to be harbouring detectable levels of BSE infectivity in any of their tissues. In considering potential human health hazards, attention should have been focused on products (other than beef) obtained from older dairy cattle that were likely to have been fed MBM. Such animals were usually slaughtered at an age where they could be close to displaying clinical signs of BSE, and therefore have infectivity in their central nervous system.

As will be discussed later, there is overwhelming evidence that vCJD is caused by the BSE agent. Studies initiated in 1996 have still failed to implicate anything other than dietary exposure to the BSE agent as the cause, but its association with any particular type of food product has not been demonstrated. One problem is that, in common with other TDEs, vCJD is likely to have long and variable incubation periods that could extend to decades. The most significant period of exposure of the UK population to the BSE agent would have been in the 1980s when the BSE epidemic was expanding rapidly, until late 1989 when the first significant measures to protect human health were introduced. At the CJD Surveillance Unit in Edinburgh, the dietary habits of individuals who became vCJD victims have been compared with controls without demonstrating any apparent significant differences. Nevertheless, there are intellectual or recall-bias problems in interviewing cases, or their relatives, regarding the eating habits of such individuals up to 20 years ago. However, it was observed that mechanically recovered meat (MRM) that was likely to have been commonly incorporated into cheap food products had featured significantly in the diets of both the vCJD cases and the controls. It was recognized that MRM was likely to have contained particles of bovine spinal cord that could contain potentially high levels of the BSE agent, and its incorporation into human foodstuff was prohibited in Great Britain in 1994. Also, when the apparently upwardly spiralling increase in the incidence of vCJD was reported in July 2000, consultants from the CJD Surveillance Unit in Edinburgh were reported in the media to have again suggested that a significant factor might have been 'institutional cuisine' in the past whereby organizations such as works canteens and school meal services may have opted to use cheaper food products that were more likely to have contained MRM. There is thus the enigma that, although respected experts appear to believe that MRM has been a potentially significant dietary vehicle whereby BSE might have been transmitted to humans to cause vCJD, this is not reflected in the studies in which the dietary habits of cases were compared with controls. However, even by July 2000, the number of cases of vCJD was not high enough to exclude chance or random factors in its aetiology. With regard to the potential consumption of MRM, the distribution of BSE infectivity in any given batch was unlikely to have beeen homogeneous. Thus, in social groups consuming food containing infected MRM at the same meal, most might have received portions that contained either no BSE infectivity, or insufficient to establish vCJD infection. In contrast, another individual might have consumed a portion of the same food that contained sufficient infectivity to have infected a number of people if the BSE infectivity had been more evenly dispersed throughout the food product. The statistical situation may also have been clouded by the operation of other 'chance factors.' For example, vCJD has only occurred thus far in individuals that are homozygous for methionine at codon 129 of their *PrP* genes. However, experience with the accidental transmission of human TDEs to humans suggests that vCJD is also likely to occur in valine homozygotes and methionine/valine heterozygotes. The simple explanation as to why vCJD has not yet apparently occurred in such genotypes is that its incubation period may be longer in such individuals. A more worrying alternative is that the clinical and neurohistopathological features of vCJD in these genotypes might not be sufficiently distinctive to differentiate it from sporadic CJD. If this proved to be the case, evidence of its occurrence might only be revealed by the relatively long-term observation of an unexpected increase in the incidence of sporadic CJD.

There are also other factors, including genetic differences at loci other than that of the *PrP* gene, that might influence the likelihood of individuals succumbing to vCJD. Also, although vCJD is clearly a neurological disease, its pathogenesis appears to involve an important earlier stage of agent replication within the lymphoreticular system.[24] Thus, if the immune system of individuals happened to be stimulated at the time of their exposure to the BSE agent, this might render them more susceptible to infection. In addition, in what appears to be a food-borne disease, lesions within the gastrointestinal tract at the time of exposure might enhance the likelihood of infection.

---

[24] A. F. Hill, R. J. Butterworth, S. Joiner, G. Jackson, M. N. Rosser, D. J. Thomas, A. Frosh, N. Tolley, J. E. Bell, M. Spencer, A. King, S. AlSarraj, J. W. Ironside, P. L. Lantos and J. Collinge, *Lancet*, 1999, **353**, 83.

The Scale of Dietary Exposure of the UK Population to the BSE Agent. Because of the considerable variety of species, including humans, that have become infected with BSE, the causal agent has frequently been considered to be particularly promiscuous. In 1988 and 1989 the first significant control measures were introduced to minimize the exposure of ruminants to infected feed, and humans to infected food, respectively. However, before the end of 1989, food was likely to have contained significant amounts of BSE infectivity. By that time, although more than 11 000 cases of BSE had been confirmed in Great Britain, it has been calculated that 446 000 infected, but apparently normal, cattle would have been processed in abattoirs to produce food.[25] Scrapie is endemic in the UK but its apparent incidence, and the quantity of sheep processed through abattoirs, is not high enough to have resulted in its contamination of the food supply with scrapie agent to anything like the extent that the BSE agent did, and no association has been found between sporadic CJD and dietary or occupational exposure to scrapie agent.[3] However, it is interesting to speculate as to whether the apparent absolute species barrier between sheep and humans (or other species) would have been so resilient if the dietary level of scrapie agent had been as high as that for the BSE agent.

4 The Evidence that vCJD is Caused by the BSE Agent

The hypothesis proposed in 1996 that vCJD resulted from exposure to the BSE agent is now supported by experimental data. It had already been demonstrated that the patterns of incubation periods in a panel of inbred strains of mice were essentially the same when injected with brain tissue from (a) British or Swiss cattle with BSE, (b) domestic cats from the UK with FSE, (c) exotic ruminants with BSE-like diseases and (d) goats, pigs and sheep that had been experimentally infected with the BSE agent. These incubation period patterns were quite unlike those for any other TDE agent. Similarly, the regional distribution and severity of spongiform changes in the brains of each of the different strains of recipient mice were the same for all of the isolates tested, but were dissimilar to those associated with other TDE agents. These studies convincingly demonstrated that the BSE agent is a single strain that has (a) affected cattle in Switzerland and the UK, (b) caused FSE, (c) infected exotic ruminants and (d) retained its distinctive phenotype after experimental passage through goats, pigs and sheep.[6] When the brain tissues from cases of vCJD were injected into the same panel of mouse strains, the same picture emerged. This was entirely different from the pattern resulting from challenge with the brain tissues from cases of sporadic CJD, and provided overwhelming evidence that the BSE agent is the cause of vCJD.[26] Confirmatory evidence was provided by the demonstration that the glycosylation patterns of the disease-specific forms of PrP derived from cattle with BSE and

[25] R. M. Anderson, C. A. Donnelly, N. M. Ferguson, M. E. Woolhouse, C. J. Watt, H. J. Udy, S. MaWhinney, S. P. Dunstan, T. R. Southwood, J. W. Wilesmith, J. B. Ryan, L. J. Hoinville, J. E. Hillerton, A. R. Austin and G. A. Wells, *Nature*, 1996, **382**, 779.

[26] M. E. Bruce, R. G. Will, J. W. Ironside, I. McConnell, D. Drummond, A. Suttie, L. McCardle, A. Chree, J. Hope, C. Birkett, S. Cousens, H. Fraser and C. J. Bostock, *Lancet*, 1997, **389**, 498.

humans with vCJD were essentially the same, but different from those derived from animals or humans with other TDEs.[27]

5 Reasons for the Occurrence of BSE in the UK and Elsewhere in Europe

It remains an open question as to whether the BSE agent was already present as a covert bovine pathogen before the UK epidemic occurred, or whether it was caused by the survival of a single thermostable strain of scrapie agent from sheep that survived the rendering processes used to manufacture MBM that was fed to cattle and other species. It is clear that the rapid expansion of the UK epidemic of BSE was due to feeding cattle with MBM that had been manufactured by the rendering industry from the tissues of BSE-infected cattle discarded mainly by abattoirs.[14]

Based upon the clinical symptoms of BSE, there have been anecdotal reports that this disease has occurred in the past. However, such reports cannot be seriously considered because UK field veterinarians with considerable experience of BSE still misdiagnose around 20% of suspects on the basis of their clinical symptoms. Also, archival studies on the brain tissue of cattle failed to reveal any neurohistopathological evidence that BSE existed in the past.[28]

Indigenous BSE does not appear to have occurred in countries that unwittingly imported infected cattle from the UK and that later developed BSE (Table 2). Thus, the epidemics that have occurred in other countries are more likely to have been precipitated through their importation of MBM from the UK, before this practice was prohibited. It is relevant that UK renderers exported more MBM into the international market in 1989 than ever before, even though the BSE risks must have been known to their customers because of the British ban on feeding ruminant-derived proteins to ruminants introduced in 1988. The international brokering system for MBM does not generally permit the identification of countries that imported UK-derived MBM during 1989 or in any other years. However, BSE has only occurred so far in European countries, and has been shown (when tested) to have been caused by the single strain of infectious agent found in Britain. This appears to constitute relatively convincing evidence that BSE became established in other European countries through their importation of UK-derived MBM before such trading was prohibited.

6 Measures Introduced to Enhance the Safety of Bovine-derived Food Products

Considering that no animal TDE had ever been shown to transmit to humans in the past, it is not surprising that the initial scientific opinion was generally that BSE was unlikely to transmit to humans but that this was not impossible. During the mid to late 1980s, the media frequently reported the opinions of self-appointed experts who proclaimed that BSE would decimate the human population, but they failed to present convincing scientific arguments to support this hypothesis.

[27] A. F. Hill, M. Desbruslais, S. Joiner, K. C. L. Sidle, I. Gowland and J. Collinge, *Nature*, 1997, **389**, 448.
[28] K. Taylor, *Vet. Rec.*, 1995, **137**, 674.

The fact that BSE eventually proved to be transmissible to humans was therefore largely a matter of serendipity rather than a foregone conclusion based upon solid scientific reasoning. Before vCJD occurred, governmental agencies and politicians had been obliged to consider a worst-case scenario in which there might be a significant risk to humans from the BSE agent. Consequently, many regulations were introduced in the UK, the EU and elsewhere to protect animal and human health at that time. Additional regulations were introduced when it became clear that there was an association between BSE and vCJD. The current regulations are complex, and only a simple overview of the effects of the most significant control measures will be considered here.

In 1988, a ban on feeding ruminant protein to ruminants was introduced in Great Britain because epidemiological studies clearly implicated the feeding of MBM to cattle as the cause of the expansion of the BSE epidemic.[14] Although cattle were usually initially fed MBM as calves, they did not generally display clinical signs of BSE until they reached an average age of five years. Thus, the benefit of the feed ban was not experienced until 1993, when a downward trend in the incidence of BSE was observed. Switzerland also introduced a ban on feeding meat and bone meal to ruminants in 1990. Even though BSE has not occurred in its cattle population, the USA also introduced a ban on feeding ruminant-derived proteins to ruminants. In 1994, EU regulations were introduced that required MBM to be manufactured using steam under pressure at 133 °C; this resulted from the early release of data produced from rendering studies involving BSE-spiked raw materials which showed that other rendering procedures were not reliable.[15] In 1994, the EU declared that all types of mammalian proteins should be excluded from ruminant diets unless it could be reliably demonstrated that ruminant protein was being excluded. After the emergence of vCJD in 1996, this ban was expanded to include the feeding of animal proteins to any form of livestock in the UK.

In late 1989, the use of specified bovine offals (SBO) in foodstuff for human consumption was prohibited in Great Britain, and in 1990 this was extended to prohibit their use in ruminant or poultry feed. At that time, nothing was known about the distribution of BSE infectivity in the tissues of infected cattle; the decision as to what should be included in the SBO list was therefore based upon what was known about the distribution of infectivity in the tissues of sheep with scrapie, as judged by mouse bioassay. The bovine tissues that were consequently excluded under the SBO regulations were the brain, spinal cord, spleen, thymus, tonsils and intestinal tract (from the duodenum to the rectum) of animals that were more than six months old. These regulations were modified in 1991 to exclude SBO regardless of the age of the cattle. Milk obtained from suspect cases was also required to be destroyed.

Later studies involving mouse bioassays of the tissues of BSE-infected cattle demonstrated that the brain and spinal cord harboured significant amounts of infectivity, but that there was no detectable infectivity in the other SBO. Because infectivity is readily detectable in the lymphoreticular tissues of sheep with scrapie, the apparent absence of infectivity in such tissues derived from BSE-infected cattle was unexpected. However, it was recognized that these negative results could have arisen because of the species barrier between cattle and mice. Studies that are still in progress have shown that the efficiency of transmitting BSE to cattle by intracerebral challenge is around 500-fold greater than transmission to mice by the same route (Wells, personal communication). However, cattle injected intracerebrally more than seven years ago with pooled spleens and lymph nodes from BSE-infected cattle remain healthy (Wells, personal communication). This is strong evidence that the lymphoreticular tissues of cattle do not become infected with the BSE agent.

With regard to milk obtained from BSE-affected cows, bioassays have failed to detect any infectivity.[29] Also, in studying the progeny of 126 beef suckler cows that developed BSE, no BSE has been observed in any of their 219 offspring that were estimated to have collectively consumed more than 111 000 litres of milk.[30]

As time progressed, a few non-SBO tissues from natural cases of BSE or experimentally infected cattle were shown to harbour infectivity. These included the retina, which resulted in the 1995 ruling that brain should be excluded from the SBO list and replaced by 'the skull, including brains and eyes.' Also, the finding of infectivity in the dorsal root ganglia that are adjacent to the spinal column led to the UK regulation introduced in 1996 (rescinded in 1999) that beef for human consumption would have to be deboned. Also in 1996, it was decreed within the EU that the export of beef and beef products from the UK (already restricted by regulations introduced in 1990 and 1994) should be completely banned. Although this decision was reversed in 1999, France and Germany were still prohibiting such imports in 2000, in contravention of EU law. Following the announcement in 1996 that vCJD was probably caused by the BSE agent, additional control measures were introduced in the UK to further protect human health. These included a selective cull of cattle considered to be most at risk of being infected, together with the offspring of BSE-infected cows. By the beginning of 2000, 76 000 at-risk cattle and 7376 offspring had been slaughtered. The OTMS (over thirty month scheme) was also introduced, whereby only cattle younger than 30 months can be used as a source of human food. By the end of 1999, 3.7 million cattle older than 30 months had been slaughtered under this scheme. Some of these carcases were sent for incineration but the remainder were rendered to produce more than 200 000 tonnes of tallow and 400 000 tonnes of meat and bone meal that have been stockpiled awaiting their safe destruction.

Within the EU, there had been concern that some member states were reluctant to accept the possibility that their cattle might be at risk of becoming infected with BSE, even when risk assessments suggested such a likelihood. Even when BSE eventually occurred in some of these states, there was a reluctance to introduce the full armoury of control measures that might be considered to be appropriate in view of the UK experience. Initially, it proved difficult to achieve any degree of agreement on the nature of legislation that might be applied throughout the EU to prevent the spread of BSE, and thus reduce any associated ruminant and human health hazards. However, agreement has now been reached which may have resulted from the fact that (a) within recent years, BSE has appeared in countries that previously appeared to be BSE-free, and (b) apart from the UK, where the incidence of BSE has been declining progressively over the

[29] D. M. Taylor, C. E. Ferguson, C. J. Bostock and M. Dawson, *Vet. Rec.*, 1995, **136**, 592.
[30] J. W. Wilesmith and J. B. M. Ryan, *Vet. Rec.*, 1992, **130**, 491.

past seven years, other countries are not experiencing the same significant downward trend in its incidence (Table 2).

The new legislation that is to be applied throughout the EU as from October 2000 requires that animal tissues considered likely to represent a BSE risk (SRM: specified risk material) should be safely discarded. These are the skull (including the brains and eyes), tonsils, spinal cord and ileum of cattle over 12 months old. Also included are the skulls (including the brains and eyes), tonsils and spinal cord of sheep and goats over 12 months old. The spleens of sheep and goats are also required to be discarded, regardless of the age of the animals.

In view of the recognized higher BSE risk in Portugal and the UK (Table 2), the prohibited materials in these countries will be the entire head (excluding the tongue but including the brains, eyes, trigeminal ganglia and tonsils), the thymus, spleen, intestine and spinal cord of cattle over six months old. If cattle are more than 30 months old, the vertebral column (including the dorsal root ganglia) must also be discarded.

There is a legal requirement within the EU that animals must be stunned before they are slaughtered in abattoirs. Stunning is intended to render animals unconscious until death ensues through exsanguination, and can be achieved by (a) the passage of electrical current through the brain, (b) the use of a gun that discharges a captive bolt that percusses against the head of the animal but does not penetrate into the brain or (c) a similarly delivered bolt that penetrates into the brain.

Penetrative stunning is sometimes accompanied by pithing but this is subject to considerable geographical variation. The pithing process involves the introduction of a flexible rod into the hole created by penetrative stunning which is prodded and agitated into the the brain and cervical spinal cord. This minimizes the involuntary movement of the animals being slaughtered before their death through exsanguination, and reduces the risk of accidental injury to abattoir personnel.

It is known that severe brain trauma in humans can result in the appearance of microscopic particles of brain tissue in the bloodstream.[31] When captive-bolts that inject air into the cranial cavity were used in cattle, relatively large particles of brain tissue were detectable in the jugular vein, heart or main pulmonary arteries of up to 33% of the cattle studied.[32-34] However, these studies did not permit any conclusions to be drawn as to whether microscopic quantities of brain tissue might traverse through the lungs into the general arterial system and eventually lodge in tissues such as muscles that are used as food. An informal survey has concluded that the compressed-air captive-bolt system is probably little used, if at all, within the EU. Nevertheless, it has been recognized that the combination of penetrative stunning and pithing might release brain tissue into the bloodstream and permit BSE infectivity to sequester in tissues destined for

[31] S. C. Ogilvy, A. C. McKee, N. J. Newman, S. M. Donnelly and K. J. Kiwak, *Neurosurgery*, 1988, **23**, 511.

[32] T. Garland, N. Bauer and M. Bailey, *Lancet*, 1996, **348**, 610.

[33] G. R. Schmidt, K. I. Hossner, R. S. Yemm and D. H. Gould, *J. Food Prot.*, 1999, **62**, 390.

[34] M. H. Anil, S. Love, S. Williams, A. Shand, J. L. McKinstry, C. R. Helps, A. Waterman-Pearson, J. Seghatchion and D. A. Harbour, *Vet. Rec.*, 1999, **145**, 460.

human consumption that would otherwise be free from infectivity. New EU legislation (operative from 31 December 2000) will prohibit the practice of pithing during the slaughter of cattle, sheep or goats intended for human consumption.

It has to be recognized that not all animal-derived constituents of human food products are obtained directly from carcases processed in abattoirs. For example, blood is collected in abattoirs and then added to food products in an untreated or treated form. Also, products such as tallow and gelatin are manufactured from tissues discarded by abattoirs, and can also be incorporated into food products.

Apart from the UK, blood collected from cattle that have been declared fit for human consumption is permitted to be included in human food products regardless of whether or not it is subjected to any further processing. Thus, untreated plasma can be found in sausages. Other blood products may be subjected to some form of heat processing before they end up in food products, but this is not required by any form of legislation. The relatively relaxed regulatory attitude to the incorporation of bovine blood into food products is largely based upon the knowledge that infectivity is generally undetetectable in the blood of animals or humans affected by TDEs. However, the potential contamination of bovine blood with the BSE agent as a result of penetrative stunning and pithing in abattoirs has been recognized and addressed, as discussed above. Clearly, if there are opportunities in the abattoir for blood to become adventitiously contaminated with brain tissue or spinal cord material, this would introduce a significant doubt concerning the safety of food products that contained blood or blood-derived products. It is considered unlikely that this could occur in the licenced abattoirs from which blood is obtained for inclusion into human food. The EU continues to monitor this situation.

Information supplied by courtesy of the Gelatin Manufacturers of Europe (GME) shows that, worldwide, the gelatin industry annually converts over one million tonnes of raw materials from animals into 220 000 tonnes of gelatin, of which 100 000 tonnes are produced in Europe. Asia, North America and South America are the next largest manufacturing areas, each with around 16% of the international market. Around 60% of the gelatin produced is used for food purposes, but much of that is produced from porcine skin. The highest use of gelatin in food products is in confectionery, followed by (on a decreasing scale) jellies, low fat and dairy products, and meat products. On a global scale, 25% of the gelatin used for food purposes is produced from bovine hides and bones. In Europe, this is somewhat lower at 15%, and not all of the bone material is sourced within Europe.

During the course of the BSE saga, the EU, FDA and other national authorities have maintained a watching brief to determine the nature of any potential problems associated with the use of gelatin in food or other products used by humans. Changing perceptions of risk have resulted in recommendations that raw materials should be obtained from areas with a low BSE risk, and that these should not contain bovine skulls or vertebral columns. Also, gelatin for human consumption is not permitted to be manufactured in the UK. Although the procedures used to manufacture gelatin are considered likely to result in a significant degree of inactivation or removal of the BSE agent, this has never been formally demonstrated. Consequently, the GME have funded a series of sophisticated validation studies to quantify this.

Although the manufacture of tallow for incorporation into foodstuff is prohibited in the UK, this practice is permitted elsewhere. It can be incorporated into a variety of products, providing that it has been manufactured from tissues declared fit for human consumption. However, this product appears to have an extremely low likelihood of being contaminated with the BSE agent, even when manufactured under worst-case conditions. Experimental rendering studies, in which abattoir waste was spiked with BSE-infected brain tissue at a level of 10%, produced tallow with no detectable infectivity, even though the most poorly inactivating rendering process produced MBM with an infectivity titre almost as high as that in the unprocessed raw materials.[15]

7 The Question as to whether BSE could be Masquerading as Scrapie in Sheep or Goats

BSE can be transmitted orally to goats and sheep by feeding them as little as 0.5 g of brain tissue from BSE-affected cattle, and the ensuing disease is clinically and neurohistopathologically indistinguishable from scrapie.[35] However, such studies are not necessarily relevant to the question as to whether goats or sheep might have become infected in the past by feeding them with MBM that was infected with the BSE agent. This is because the infectivity titre in MBM could not have been nearly as high as that in the BSE-infected brain tissue used in the oral transmission studies. It is difficult to resolve whether transmission of BSE to sheep or goats might have occurred in countries within which scrapie, but not BSE in cattle, is endemic, but where they might have been fed MBM imported from the UK. The situation is even more complicated in countries where both endemic BSE and scrapie occur if sheep or goats were fed MBM. In the UK, where such a situation exists, there are a number of ongoing studies designed to determine whether the BSE agent has infected sheep, but these may not produce unequivocal data. A further clue might be provided by strain typing some of the agents that have caused singleton cases of scrapie in 12 different Swiss sheep flocks during the 1990s. Such a study could be relevant because Switzerland appeared to be scrapie-free before the emergence of its BSE epidemic, and it is already known that the strain type of the BSE agent in Switzerland is the same as that of the British BSE agent. Although it would be considered unusual for only single sheep to become scrapie affected in all of these Swiss flocks, the occurrence of single, or few, cases of BSE in herds of cattle that were fed potentially BSE-infected MBM is not unusual. This is considered to have resulted from the uneven distribution of BSE infectivity throughout any given batch of MBM by virtue of the manufacturing process.[36]

8 Future Trends

Since 1993, the year-to-year incidence of BSE in the UK has shown a significant

[35] J. D. Foster, J. Hope and H. Fraser, *Vet. Rec.*, 1993, **133**, 339.
[36] D. M. Taylor, *Transcripts of an International Conference on Meat and Bone Meal*, Brussels, 1997.

decline as a result of the introduction of progressively rigorous control measures, but it still remains to be seen whether or not complete eradication can be achieved, However, the same consistent downward trend has not been experienced in other European countries within which endemic BSE occurs (Table 2). It remains an open question as to what the eventual scale of the epidemics, and the commensurate risks to human health, will be in these countries. As has been discussed, the EU has introduced new measures to minimize the expansion of the incidence of BSE in countries already affected, and to avoid the spread of this disease to other countries. However, the UK experience has been that regulatory measures are only effective if they are adequately monitored. This is reflected in the current UK position whereby there are frequent visits by government inspectors to abattoirs, knackers, hunt kennels, rendering plants, feed mills and incineration facilities to ensure compliance with all of the relevant regulations. There is also a strict policy that, where there is sufficient evidence, those that have broken the law will be prosecuted unless their transgressions were of an innocent nature. The UK has learned these lessons from bitter experience, but it remains to be seen whether or not other European countries accept the need, or are prepared, to commit such considerable resources to tackle their BSE problems.

Natural and Synthetic Chemicals in the Diet: a Critical Analysis of Possible Cancer Hazards

LOIS SWIRSKY GOLD, THOMAS H. SLONE AND BRUCE
N. AMES

1 Introduction

Possible cancer hazards in food have been much discussed and hotly debated in the scientific literature, the popular press, the political arena, and the courts. Consumer opinion surveys indicate that much of the U.S. public believes that pesticide residues in food are a serious cancer hazard. In contrast, epidemiological studies indicate that the major preventable risk factors for cancer are smoking, dietary imbalances, endogenous hormones, and inflammation, *e.g.* from chronic infections. Other important factors include intense sun exposure, lack of physical activity, and excess alcohol consumption.[1] Overall cancer death rates in the U.S. (excluding lung cancer due to smoking) have declined 19% since 1950[2]. The types of cancer deaths that have decreased since 1950 are primarily stomach, cervical, uterine, and colorectal. The types that have increased are primarily lung cancer (87% is due to smoking, as are 31% of all cancer deaths in the U.S.[3]), melanoma (probably due to sunburns), and non-Hodgkin's lymphoma. If lung cancer is included, mortality rates have increased over time, but recently have declined.[2]

Thus, epidemiological studies do not support the idea that synthetic pesticide residues are important for human cancer. Although some epidemiological studies find an association between cancer and low levels of some industrial pollutants, the studies often have weak or inconsistent results, rely on ecological correlations

---

[1] B.N. Ames, L.S. Gold and W.C. Willett, *Proc. Natl. Acad. Sci. USA*, 1995, **92**, 5258; http://socrates.berkeley.edu/mutagen/ames.pnas3.html.
[2] L.A.G. Ries, M.P. Eisner, C.L. Kosary, B.F. Hankey, B.A. Miller, L. Clegg and B.K. Edwards (eds.), *SEER Cancer Statistics Review, 1973–1997*, National Cancer Institute, Bethesda, 2001.
[3] American Cancer Society, *Cancer Facts & Figures—2000*, American Cancer Society, Atlanta, 2000.

Issues in Environmental Science and Technology No. 15
Food Safety and Food Quality

or indirect exposure assessments, use small sample sizes, and do not control for confounding factors such as composition of the diet, which is a potentially important confounder. Outside the workplace, the levels of exposure to synthetic pollutants or pesticide residues are low and rarely seem toxicologically plausible as a causal factor when compared to the wide variety of naturally occurring chemicals to which all people are exposed.[4] Whereas public perceptions tend to identify *chemicals* as being only synthetic and only synthetic chemicals as being toxic, every natural chemical is also toxic at some dose, and the vast proportion of chemicals to which humans are exposed are naturally occurring (see Section 2).

There is a paradox in the public concern about possible cancer hazards from pesticide residues in food and the lack of public understanding of the substantial evidence indicating that high consumption of the foods which contain pesticide residues—fruits and vegetables—has a protective effect against many types of cancer. A review of about 200 epidemiological studies reported a consistent association between low consumption of fruits and vegetables and cancer incidence at many target sites.[5] The quarter of the population with the lowest dietary intake of fruits and vegetables has roughly twice the cancer rate for many types of cancer (lung, larynx, oral cavity, esophagus, stomach, colon and rectum, bladder, pancreas, cervix, and ovary) compared to the quarter with the highest consumption of those foods. The protective effect of consuming fruits and vegetables is weaker and less consistent for hormonally related cancers, such as breast cancer. Studies suggest that inadequate intake of many micronutrients in these foods may be radiation mimics and are important in the protective effect.[6] Despite the substantial evidence of the importance of fruits and vegetables in prevention, half the American public did not identify fruit and vegetable consumption as a protective factor against cancer.[7] Consumption surveys, moreover, indicate that 80% of children and adolescents in the U.S.[8] and 68% of adults[9] did not consume the intake of fruits and vegetables recommended by the National Cancer Institute and the National Research Council: five servings per day. One important consequence of inadequate consumption of fruits and vegetables is low intake of some micronutrients. For example, folic acid is one of the most common vitamin deficiencies in the population consuming few dietary fruits and vegetables; folate deficiency causes chromosome breaks in humans by a mechanism that mimics radiation.[6] Approximately 10% of the U.S. population[10] had a lower folate level than that at which chromosome breaks occur. Folate supplementation above the RDA minimized chromosome breakage.[11]

[4] L.S. Gold, T.H. Slone, B.R. Stern, N.B. Manley and B.N. Ames, *Science*, 1992, **258**, 261; http://potency.berkeley.edu/text/science.html.
[5] G. Block, B. Patterson and A. Subar, *Nutr. Cancer*, 1992, **18**, 1.
[6] B.N. Ames, *Toxicol. Lett.*, 1998, **103**, 5.
[7] U.S. National Cancer Institute, *J. Natl. Cancer Inst.*, 1996, **88**, 1314.
[8] S.M. Krebs-Smith, A. Cook, A.F. Subar, L. Cleveland, J. Friday and L.L. Kahle, *Arch. Pediatr. Adolesc. Med.*, 1996, **150**, 81.
[9] S.M. Krebs-Smith, A. Cook, A.F. Subar, L. Cleveland and J. Friday, *Am. J. Public Health*, 1995, **85**, 1623.
[10] L.S. Gold, B.N. Ames and T.H. Slone, in *Human and Environmental Risk Assessment: Theory and Practice*, ed. D. Paustenbach, Wiley, New York, in press.
[11] M. Fenech, C. Aitken and J. Rinaldi, *Carcinogenesis*, 1998, **19**, 1163.

Given the lack of epidemiological evidence to link dietary synthetic pesticide residues to human cancer, and taking into account public concerns about pesticide residues as possible cancer hazards, public policy with respect to pesticides has relied on the results of high-dose, rodent cancer tests as the major source of information for assessing potential cancer risks to humans. This article examines critically the assumptions, methodology, results, and implications of cancer risk assessments of pesticide residues in the diet and compares results for synthetic pesticides to results for naturally occurring chemicals in food. Our analyses are based on results in our Carcinogenic Potency Database (CPDB),[12,13] which provide the necessary data to examine the published literature of chronic animal cancer tests; the CPDB includes results of 5620 experiments on 1372 chemicals. Specifically, the following are addressed:

- Human exposure to synthetic pesticide residues compared to the broader and greater exposure to natural chemicals in the diet
- Cancer risk assessment methodology, including the use of animal data from high-dose bioassays in which half the chemicals tested are carcinogenic
- Increased cell division as an important hypothesis for the high positivity rate in rodent bioassays, and the implications for risk assessment
- Providing a broad perspective on possible cancer hazards from a variety of exposures to rodent carcinogens, including natural dietary chemicals and synthetic chemicals, by ranking on the HERP index: Human Exposure/Rodent Potency
- Identification and ranking of exposures in the U.S. diet to naturally occurring chemicals that have not been tested for carcinogenicity, using an index that takes into account the toxic dose of a chemical (LD_{50}) and average consumption in the U.S. diet.

2   Human Exposures to Natural and Synthetic Chemicals

Current regulatory policy to reduce human cancer risks is based on the idea that chemicals which induce tumors in rodent cancer bioassays are potential human carcinogens. The chemicals selected for testing in rodents, however, are primarily synthetic.[12,13] The enormous background of human exposures to natural chemicals has not been systematically examined. This has led to an imbalance in both data and perception about possible carcinogenic hazards to humans from chemical exposures. The regulatory process does not take into account: (1) that natural chemicals make up the vast bulk of chemicals to which humans are exposed; (2) that the toxicology of synthetic and natural toxins is not fundamentally different; (3) that about half of the chemicals tested, whether natural or synthetic, are carcinogens when tested using current experimental protocols; (4) that testing for carcinogenicity at near-toxic doses in rodents does not provide enough information to predict the excess number of human cancers that might occur at low-dose exposures; (5) that testing at the maximum tolerated dose (MTD)

[12] L. S. Gold and E. Zeiger (eds.), *Handbook of Carcinogenic Potency and Genotoxicity Databases*, CRC Press, Boca Raton, 1997; http://potency.berkeley.edu/crcbook.html.

[13] L. S. Gold, N. B. Manley, T. H. Slone and L. Rohrbach, *Environ. Health Perspect.*, 1999, **107** (Suppl. 4), 527; http://ehpnet1.niehs.nih.gov/docs/1999/suppl-4/toc.html.

frequently can cause chronic cell killing and consequent cell replacement (a risk factor for cancer that can be limited to high doses), and that ignoring this effect in risk assessment can greatly exaggerate risks.

We estimate that about 99.9% of the chemicals that humans ingest are natural. The amounts of synthetic pesticide residues in plant foods are low in comparison to the amount of natural pesticides produced by plants themselves.[14,15] Of all dietary pesticides that Americans eat, 99.99% are natural: they are the chemicals produced by plants to defend themselves against fungi, insects, and other animal predators.[14,15] Each plant produces a different array of such chemicals.

We estimate that the daily average American exposure to natural pesticides in the diet is about 1500 mg and to burnt material is about 2000 mg.[15] In comparison, the total daily exposure to all synthetic pesticide residues combined is about 0.09 mg, based on the sum of residues reported by the U.S. Food and Drug Administration (FDA) in its study of the 200 synthetic pesticide residues thought to be of greatest concern.[16,17] Humans ingest roughly 5000 to 10 000 different natural pesticides and their breakdown products.[14] Despite this enormously greater exposure to natural chemicals, among the chemicals tested in long-term bioassays in the CPDB, 77% (1051/1373) are synthetic (i.e. do not occur naturally).[12,13]

Concentrations of natural pesticides in plants are usually measured in parts per thousand or million rather than parts per billion, which is the usual concentration of synthetic pesticide residues. Therefore, since humans are exposed to so many more natural than synthetic chemicals (by weight and by number), human exposure to natural rodent carcinogens, as defined by high-dose rodent tests, is ubiquitous.[14] It is probable that almost every fruit and vegetable in the supermarket contains natural pesticides that are rodent carcinogens. Even though only a tiny proportion of natural pesticides have been tested for carcinogenicity, 37 of 71 that have been tested are rodent carcinogens that are present in the common foods listed in Table 1.

Humans also ingest numerous natural chemicals that are produced as by-products of cooking food. For example, more than 1000 chemicals have been identified in roasted coffee, many of which are produced by roasting.[4] Only 30 have been tested for carcinogenicity according to the most recent results in our CPDB, and 21 of these are positive in at least one test (Table 2) totaling at least 10 mg of rodent carcinogens per cup. Among the rodent carcinogens in coffee are the plant pesticides caffeic acid (present at 1800 ppm) and catechol (present at 100 ppm). Two other plant pesticides in coffee, chlorogenic acid and neochlorogenic acid (present at 21 600 ppm and 11 600 ppm, respectively) are metabolized to caffeic acid and catechol but have not been tested for carcinogenicity. Chlorogenic acid and caffeic acid are mutagenic,[14] and clastogenic. For another plant

[14] B. N. Ames, M. Profet and L. S. Gold, *Proc. Natl. Acad. Sci. USA*, 1990, **87**, 7777; http://socrates.berkeley.edu/mutagen/PNAS2.html.
[15] B. N. Ames, M. Profet and L. S. Gold, *Proc. Natl. Acad. Sci. USA*, 1990, **87**, 7782.
[16] L. S. Gold, T. H. Slone and B. N. Ames, in *Handbook of Pesticide Toxicology*, ed. R. I. Krieger, Academic Press, New York, in press.
[17] L. S. Gold, B. R. Stern, T. H. Slone, J. P. Brown, N. B. Manley and B. N. Ames, *Cancer Lett.*, 1997, **117**, 195; http://potency.berkeley.edu/text/pesticide.html.

Table 1 Carcinogenicity status of natural pesticides tested in rodents[a]	**Carcinogens:**[b] $N = 37$ — Acetaldehyde methylformylhydrazone, allyl isothiocyanate, arecoline.HCl, benzaldehyde, benzyl acetate, caffeic acid, capsaicin, catechol, clivorine, coumarin, crotonaldehyde, 3,4-dihydrocoumarin, estragole, ethyl acrylate, N2-γ-glutamyl-p-hydrazino-benzoic acid, hexanal methylformylhydrazone, p-hydrazinobenzoic acid.HCl, hydroquinone, 1-hydroxyanthraquinone, lasiocarpine, d-limonene, 3-methoxycatechol, 8-methoxypsoralen, N-methyl-N-formylhydrazine, α-methylbenzyl alcohol, 3-methylbutanal methylformylhydrazone, 4-methylcatechol, methylhydrazine, monocrotaline, pentanal methylformylhydrazone, petasitenine, quercetin, reserpine, safrole, senkirkine, sesamol, symphytine
	Non-carcinogens: $N = 34$ — Atropine, benzyl alcohol, benzyl isothiocyanate, benzyl thiocyanate, biphenyl, d-carvone, codeine, deserpidine, disodium glycyrrhizinate, ephedrine sulfate, epigallo-catechin, eucalyptol, eugenol, gallic acid, geranyl acetate, β-N-[γ-l(+)-glutamyl]-4-hydroxymethylphenylhydrazine, glycyrrhetinic acid, p-hydrazinobenzoic acid, isosafrole, kaempferol, dl-menthol, nicotine, norharman, phenethyl isothiocyanate, pilocarpine, piperidine, protocatechuic acid, rotenone, rutin sulfate, sodium benzoate, tannic acid, 1-$trans$-δ^9-tetrahydrocannabinol, turmeric oleoresin, vinblastine

[a]Fungal toxins are not included.
[b]These rodent carcinogens occur in: absinthe, allspice, anise, apple, apricot, banana, basil, beet, broccoli, Brussels sprouts, cabbage, cantaloupe, caraway, cardamom, carrot, cauliflower, celery, cherries, chili pepper, chocolate, cinnamon, cloves, coffee, collard greens, comfrey herb tea, corn, coriander, currants, dill, eggplant, endive, fennel, garlic, grapefruit, grapes, guava, honey, honeydew melon, horseradish, kale, lemon, lentils, lettuce, licorice, lime, mace, mango, marjoram, mint, mushrooms, mustard, nutmeg, onion, orange, paprika, parsley, parsnip, peach, pear, peas, black pepper, pineapple, plum, potato, radish, raspberries, rhubarb, rosemary, rutabaga, sage, savory, sesame seeds, soybean, star anise, tarragon, tea, thyme, tomato, turmeric, and turnip.

pesticide in coffee, d-limonene, the only tumors induced were in male rat kidney, by a mechanism involving accumulation of α_{2u}-globulin and increased cell division in the kidney, which would not be predictive of a carcinogenic hazard to humans.[18] Some other rodent carcinogens in coffee are products of cooking, *e.g.* furfural and benzo[a]pyrene. The point here is not to indicate that rodent data necessarily implicate coffee as a risk factor for human cancer, but rather to illustrate that there is an enormous background of chemicals in the diet that are natural and that have not been a focus of carcinogenicity testing. A diet free of naturally occurring chemicals that are carcinogens in high-dose rodent tests is impossible.

It is often assumed that because natural chemicals are part of human

[18] International Agency for Research on Cancer, *IARC Monographs on the Evaluation of Carcinogenic Risk of Chemicals to Humans*, IARC, Lyon, 1971–1999, vols. 1–73, Suppl. 7.

Table 2 Carcinogenicity in rodents of natural chemicals in roasted coffee

Positive: N = 21	Acetaldehyde, benzaldehyde, benzene, benzofuran, benzo[*a*]pyrene, caffeic acid, catechol, 1,2,5,6-dibenzanthracene, ethanol, ethylbenzene, formaldehyde, furan, furfural, hydrogen peroxide, hydroquinone, isoprene, limonene, 4-methylcatechol, styrene, toluene, xylene
Not positive: N = 8	Acrolein, biphenyl, choline, eugenol, nicotinamide, nicotinic acid, phenol, piperidine
Uncertain:	Caffeine
Yet to test:	∼1000 chemicals

evolutionary history, whereas synthetic chemicals are recent, the mechanisms that have evolved in animals to cope with the toxicity of natural chemicals will fail to protect against synthetic chemicals, including synthetic pesticides. This assumption is flawed for several reasons:

1. Humans have many natural defenses that buffer against normal exposures to toxins[15] and these are usually general, rather than tailored for each specific chemical. Thus they work against both natural and synthetic chemicals. Examples of general defenses include the continuous shedding of cells exposed to toxins—the surface layers of the mouth, esophagus, stomach, intestine, colon, skin, and lungs are discarded every few days; DNA repair enzymes, which repair DNA that was damaged from many different sources; and detoxification enzymes of the liver and other organs which generally target classes of chemicals rather than individual chemicals. That human defenses are usually general, rather than specific for each chemical, makes good evolutionary sense. The reason that predators of plants evolved general defenses is presumably to be prepared to counter a diverse and ever-changing array of plant toxins in an evolving world; if a herbivore had defenses against only a specific set of toxins, it would be at great disadvantage in obtaining new food when favored foods became scarce or evolved new chemical defenses.

2. Various natural toxins, which have been present throughout vertebrate evolutionary history, nevertheless cause cancer in vertebrates. Mold toxins, such as aflatoxin, have been shown to cause cancer in rodents, monkeys, humans, and other species. Many of the common elements, despite their presence throughout evolution, are carcinogenic to humans at high doses, *e.g.* salts of cadmium, beryllium, nickel, chromium, and arsenic. Furthermore, epidemiological studies from various parts of the world indicate that certain natural chemicals in food may be carcinogenic risks to humans; for example, the chewing of betel nut with tobacco is associated with oral cancer. Among the agents identified as human carcinogens by the International Agency for Research in Cancer (IARC), 65% (39/60) occur naturally: 15 are natural chemicals, 11 are mixtures of natural chemicals, and 5 are infectious agents.[18] Thus, the idea that a chemical is 'safe', because it is natural, is not correct.

3. Humans have not had time to evolve a 'toxic harmony' with all of their dietary plants. The human diet has changed markedly in the last few thousand years. Indeed, very few of the plants that humans eat today, *e.g.* coffee, cocoa, tea, potatoes, tomatoes, corn, avocados, mangos, olives, and kiwi fruit, would have

been present in a hunter-gatherer's diet. Natural selection works far too slowly for humans to have evolved specific resistance to the food toxins in these newly introduced plants.

4. Some early synthetic pesticides were lipophilic organochlorines that persist in nature and bioaccumulate in adipose tissue, *e.g.* DDT, aldrin, dieldrin (DDT is discussed in Section 5). This ability to bioaccumulate is often seen as a hazardous property of synthetic pesticides; however, such bioconcentration and persistence are properties of relatively few synthetic pesticides. Moreover, many thousands of chlorinated chemicals are produced in nature.[19] Natural pesticides also can bioconcentrate if they are fat soluble. Potatoes, for example, were introduced into the worldwide food supply a few hundred years ago; potatoes contain solanine and chaconine, which are fat-soluble, neurotoxic, natural pesticides that can be detected in the blood of all potato-eaters. High levels of these potato neurotoxins have been shown to cause birth defects in rodents.[16]

5. Since no plot of land is free from attack by insects, plants need chemical defenses—either natural or synthetic—to survive pest attack. Thus, there is a trade-off between naturally occurring pesticides and synthetic pesticides. One consequence of efforts to reduce pesticide use is that some plant breeders develop plants to be more insect-resistant by making them higher in natural pesticides. A recent case illustrates the potential hazards of this approach to pest control: when a major grower introduced a new variety of highly insect-resistant celery into commerce, people who handled the celery developed rashes when they were subsequently exposed to sunlight. Some detective work found that the pest-resistant celery contained 6200 parts per billion (ppb) of carcinogenic (and mutagenic) psoralens instead of the 800 ppb present in common celery.[20]

3 The High Carcinogenicity Rate among Chemicals Tested in Chronic Animal Cancer Tests

Since the toxicology of natural and synthetic chemicals is similar, one expects, and finds, a similar positivity-rate for carcinogenicity among synthetic and natural chemicals that have been tested in rodent bioassays. Among chemicals tested in rats and mice in the CPDB, about half the natural chemicals are positive, and about half of all chemicals tested are positive. This high positivity rate in rodent carcinogenesis bioassays is consistent for many data sets (Table 3): among chemicals tested in rats and mice, 59% (350/590) are positive in at least one experiment, 60% of synthetic chemicals (271/451), and 57% of naturally occurring chemicals (79/139). Among chemicals tested in at least one species, 52% of natural pesticides (37/71) are positive, 61% of fungal toxins (14/23), and 70% of the chemicals in roasted coffee (21/30) (Table 2). Among commercial pesticides reviewed by the U.S. EPA[21] the positivity rate is 41% (79/194); this rate is similar among commercial pesticides that also occur naturally and those that

[19] G. W. Gribble, *Pure Appl. Chem.*, 1996, **68**, 1699.
[20] R. C. Beier and H. N. Nigg, in *Foodborne Disease Handbook*, ed. Y. H. Hui, J. R. Gorham, K. D. Murrell and D. O. Cliver, Dekker, New York, 1994, pp. 1–186.
[21] U.S. Environmental Protection Agency, *Status of Pesticides in Registration, Reregistration, and Special Review*, USEPA, Washington, 1998.

Table 3 Proportion of chemicals evaluated as carcinogenic[a]

Chemicals tested in both rats and mice	
Chemicals in the CPDB	350/590 (59%)
Naturally occurring chemicals in the CPDB	79/139 (57%)
Synthetic chemicals in the CPDB	271/451 (60%)
Chemicals tested in rats and/or mice	
Chemicals in the CPDB	702/1348 (52%)
Natural pesticides in the CPDB	37/71 (52%)
Mold toxins in the CPDB	14/23 (61%)
Chemicals in roasted coffee in the CPDB	21/30 (70%)
Commercial pesticides	79/194 (41%)
Innes negative chemicals retested	17/34 (50%)
Physician's Desk Reference (PDR): drugs with reported cancer tests[b]	117/241 (49%)
FDA database of drug submissions[b]	125/282 (44%)

[a]From the Carcinogenic Potency Database.[12,13]
[b]140 drugs are in both the FDA and PDR databases.

are only synthetic, as well as between commercial pesticides that have been cancelled and those still in use.

Since the results of high-dose rodent tests are routinely used to identify a chemical as a possible cancer hazard to humans, it is important to try to understand how representative the 50% positivity rate might be of all untested chemicals. If half of all chemicals (both natural and synthetic) to which humans are exposed would be positive if tested, then the utility of a test to identify a chemical as a 'potential human carcinogen' because of an increase in tumor incidence is questionable. To determine the true proportion of rodent carcinogens among chemicals would require a comparison of a random group of synthetic chemicals to a random group of natural chemicals. Such an analysis has not been done.

It has been argued that the high positivity rate is due to selecting more suspicious chemicals to test for carcinogenicity. For example, chemicals may be selected that are structurally similar to known carcinogens or genotoxins. That is a likely bias since cancer testing is both expensive and time-consuming, making it prudent to test suspicious compounds. On the other hand, chemicals are selected for testing for many reasons, including the extent of human exposure, level of production, and scientific questions about carcinogenesis. Among chemicals tested in both rats and mice, mutagens are positive in rodent bioassays more frequently than non-mutagens: 80% of mutagens are positive (176/219) compared to 50% (135/271) of non-mutagens. Thus, if testing is based on suspicion of carcinogenicity, then more mutagens should be selected than non-mutagens; however, of the chemicals tested in both species, 55% (271/490) are not mutagenic. This suggests that prediction of positivity is often not the basis for selecting a chemical to test. Another argument against selection bias is the high positivity rate for drugs (Table 3), because drug development tends to favor chemicals that are not mutagens or suspected carcinogens. In the *Physician's Desk Reference* (PDR), however, 49% (117/241) of the drugs that report results of animal cancer tests are carcinogenic (Table 3).[22]

Table 4 Results of subsequent tests on chemicals (primarily pesticides) not found carcinogenic by Innes *et al.* (1968)[a]

Retested chemicals	% Carcinogenic when retested		
	Mice	Rats	Either mice or rats
All retested	7/26 (27%)	14/34 (41%)	17/34 (50%)
Innes: not carcinogenic	3/10 (30%)	9/18 (50%)	10/18 (56%)
Innes: needs further evaluation	4/16 (25%)	5/16 (31%)	7/16 (44%)

[a]Of 119 chemicals tested by Innes *et al.*, 11 (9%) were evaluated as positive by Innes *et al.* (M) = positive in mice when retested; (R) = positive in rats when retested.

Carcinogenic when retested: atrazine (R), azobenzene (R),[b] captan (M, R), carbaryl (R), 3-(*p*-chlorophenyl)-1,1-dimethylurea (R),[b] *p,p′*-DDD (M),[b] folpet (M), manganese ethylenebisthiocarbamate (R), 2-mercaptobenzothiazole (R), *N*-nitrosodiphenylamine (R),[b] 2,3,4,5,6-pentachlorophenol (M, R), *o*-phenylphenol (R), piperonyl butoxide (M, R),[b] piperonyl sulfoxide (M),[b] 2,4,6-trichlorophenol (M, R),[b] zinc dimethyldithiocarbamate (R), zinc ethylenebisthiocarbamate (R).

Not carcinogenic when retested: (2-chloroethyl)trimethylammonium chloride,[b] calcium cyanamide,[b] diphenyl-*p*-phenylenediamine, endosulfan, *p,p′*-ethyl-DDD,[b] ethyl tellurac,[b] isopropyl *N*-(3-chlorophenyl)carbamate, lead dimethyldithiocarbamate,[b] maleic hydrazide, mexacarbate,[b] monochloroacetic acid, phenyl-β-naphthylamine,[b] rotenone, sodium diethyldithiocarbamate trihydrate,[b] tetraethylthiuram disulfide,[b] tetramethylthiuram disulfide, 2,4,5-trichlorophenoxyacetic acid.

[b] = Innes *et al.* stated that further testing was needed.

Moreover, while some chemical classes are more often carcinogenic in rodent bioassays than others, *e.g.* nitroso compounds, aromatic amines, nitroaromatics, and chlorinated compounds, prediction is still imperfect. For example, a prospective prediction exercise was conducted by several experts in 1990 in advance of the two-year NTP bioassays. There was wide disagreement among the experts on which chemicals would be carcinogenic when tested and the level of accuracy varied by expert, thus indicating that predictive knowledge is uncertain.[23]

One large series of mouse experiments by Innes *et al.*[24] has been frequently cited as evidence that the true proportion of rodent carcinogens is actually low among tested substances (Table 4). In the Innes study, 119 synthetic pesticides and industrial chemicals were tested, and only 11 (9%) were evaluated as carcinogenic. Our analysis indicates that those early experiments lacked power to detect an effect because they were conducted only in mice (not in rats), they included only 18 animals in a group (compared with the standard protocol of 50), the animals were tested for only 18 months (compared with the standard 24 months), and the Innes dose was usually lower than the highest dose in subsequent mouse tests if the same chemical was tested again.[12,13]

To assess whether the low positivity rate in the Innes study was due to the lack of power in the design of the experiments, we used results in our CPDB to examine subsequent bioassays on the Innes chemicals that had not been evaluated as positive (results and chemical names are reported in Table 4).

[22] T. S. Davies and A. Monro, *J. Am. Coll. Toxicol.*, 1995, **14**, 90.

[23] G. S. Omenn, S. Stuebbe and L. B. Lave, *Mol. Carcinog.*, 1995, **14**, 37.

[24] J. R. M. Innes, B. M. Ulland, M. G. Valerio, L. Petrucelli, L. Fishbein, E. R. Hart, A. J. Pallota, R. R. Bates, H. L. Falk, J. J. Gart, M. Klein, I. Mitchell and J. Peters, *J. Natl. Cancer Inst.*, 1969, **42**, 1101.

Among the 34 chemicals that were not positive in the Innes study and were subsequently retested with more standard protocols, 17 had a subsequent positive evaluation of carcinogenicity (50%), which is similar to the proportion among all chemicals in the CPDB (Table 4). Of the 17 new positives, 7 were carcinogenic in mice and 14 in rats. Innes *et al.* had recommended further evaluation of some chemicals that had inconclusive results in their study. If those were the chemicals subsequently retested, then one might argue that they would be the most likely to be positive. Our analysis does not support that view, however. We found that the positivity rate among the chemicals that the Innes study said needed further evaluation was 7 of 16 (44%) when retested, compared to 10 of 18 (56%) among the chemicals that Innes evaluated as negative. Our analysis thus supports the idea that the low positivity rate in the Innes study of synthetic pesticides and pollutants resulted from lack of power.

Since many of the chemicals tested by Innes *et al.* were synthetic pesticides, we re-examined the question of what proportion of synthetic pesticides are carcinogenic (as shown in Table 3) by excluding the pesticides tested only in the Innes series. The Innes studies had little effect on the positivity rate: Table 3 indicates that of all commercial pesticides in the CPDB, 41% (79/194) are rodent carcinogens; when the analysis is repeated by excluding the chemicals tested only with the Innes protocol, 47% (77/165) are carcinogens.

4 The Importance of Cell Division in Mutagenesis and Carcinogenesis

What might explain the high positivity rate among chemicals tested in rodent cancer bioassays (Table 3)? In standard cancer tests, rodents are given a chronic, near-toxic dose: the maximum tolerated dose (MTD). Evidence is accumulating that cell division caused by the high dose itself, rather than the chemical *per se*, contributes to cancer in such tests.[25-27] High doses can cause chronic wounding of tissues, cell death, and consequent chronic cell division of neighboring cells, which is a risk factor for cancer.[25] Each time a cell divides, there is some probability that a mutation will occur, and thus increased cell division increases the risk of cancer. At the low levels of pesticide residues to which humans are usually exposed, such increased cell division does not occur. The process of mutagenesis and carcinogenesis is complicated because many factors are involved, *e.g.* DNA lesions, DNA repair, cell division, clonal instability, apoptosis, and p53 (a cell cycle control gene that is mutated in half of human tumors).[28] The normal endogenous level of oxidative DNA lesions in somatic cells is appreciable.[29] In addition, tissues injured by high doses of chemicals have an inflammatory immune response involving activation of white cells in response

[25] B. N. Ames and L. S. Gold, *Proc. Natl. Acad. Sci. USA*, 1990, **87**, 7772; http://socrates. berkeley.edu/mutagen/PNAS1.html.

[26] S. M. Cohen, *Drug Metab. Rev.*, 1998, **30**, 339.

[27] B. E. Butterworth and M. S. Bogdanffy, *Regul. Toxicol. Pharmacol.*, 1999, **29**, 23.

[28] J. G. Christensen, T. L. Goldsworthy and R. C. Cattley, *Mol. Carcinog.*, 1999, **25**, 273.

[29] H. J. Helbock, K. B. Beckman, M. K. Shigenaga, P. B. Walter, A. A. Woodall, H. C. Yeo and B. N. Ames, *Proc. Natl. Acad. Sci. USA*, 1998, **95**, 288.

to cell death.[30] Activated white cells release mutagenic oxidants (including peroxynitrite, hypochlorite, and H_2O_2). Therefore, the very low levels of chemicals to which humans are exposed through water pollution or synthetic pesticide residues may pose no or only minimal cancer risks.

It seems likely that a high proportion of all chemicals, whether synthetic or natural, might be 'carcinogens' if administered in the standard rodent bioassay at the MTD, primarily due to the effects of high doses on cell division and DNA damage.[25,27,31] For non-mutagens, cell division at the MTD can increase carcinogenicity, and for mutagens there can be a synergistic effect between DNA damage and cell division at high doses. *Ad libitum* feeding in the standard bioassay can also contribute to the high positivity rate;[32] in calorie-restricted mice, cell division rates are markedly lowered in several tissues. Without additional data on how a chemical causes cancer, the interpretation of a positive result in a rodent bioassay is highly uncertain.

Although cell division is not measured in routine cancer tests, many studies on rodent carcinogenicity show a correlation between cell division at the MTD and cancer.[25] Extensive reviews of bioassay results document that chronic cell division can induce cancer.[12,33,34] A large epidemiological literature[35] indicates that increased cell division by hormones and other agents can increase human cancer.

Several of our findings in large-scale analyses of the results of animal cancer tests,[12,36] are consistent with the idea that cell division increases the carcinogenic effect in high dose bioassays, including the high proportion of chemicals that are positive; the high proportion of rodent carcinogens that are not mutagenic; the fact that mutagens, which can both damage DNA and increase cell division at high doses, are more likely than non-mutagens to be positive, to induce tumors in both rats and mice, and to induce tumors at multiple sites.[12,36] Analyses of the limited data on dose–response in bioassays are consistent with the idea that cell division from cell killing and cell replacement is important. Among rodent bioassays with two doses and a control group, about half the tumor incidence rates that are evaluated as target sites are statistically significant at the MTD but not at half the MTD ($p < 0.05$). The proportions are similar for mutagens (44%, 148/334) and non-mutagens (47%, 76/163),[12,13] suggesting that cell division at the MTD may be important for the carcinogenic response of mutagens as well as non-mutagens that are rodent carcinogens.

To the extent that increases in tumor incidence in rodent studies are due to the secondary effects of inducing cell division at the MTD, then any chemical is a likely rodent carcinogen, and carcinogenic effects can be limited to high doses. Linearity of the dose–response also seems less likely than has been assumed

30 R. A. Roberts and I. Kimber, *Carcinogenesis*, 1999, **20**, 1397.

31 M. L. Cunningham and H. B. Matthews, *Toxicol. Appl. Pharmacol.*, 1991, **110**, 505.

32 R. Hart, D. Neumann and R. Robertson, *Dietary Restriction: Implications for the Design and Interpretation of Toxicity and Carcinogenicity Studies*, ILSI Press, Washington, 1995.

33 J. L. Counts and J. I. Goodman, *Regul. Toxicol. Pharmacol.*, 1995, **21**, 418.

34 L. S. Gold, T. H. Slone and B. N. Ames, *Drug Metab. Rev.*, 1998, **30**, 359; http://potency.berkeley.edu/text/drugmetrev.html.

35 S. Preston-Martin, M. C. Pike, R. K. Ross and P. A. Jones, *Cancer Res.*, 1990, **50**, 7415.

36 L. S. Gold, T. H. Slone, B. R. Stern and L. Bernstein, *Mutat. Res.*, 1993, **286**, 75.

L. Swirsky Gold, T. H. Slone and B. N. Ames

because of the inducibility of numerous defense enzymes which deal with exogenous chemicals as groups, *e.g.* oxidants, electrophiles, and thus protect humans against natural and synthetic chemicals, including potentially mutagenic reactive chemicals.[15] Thus, true risks at the low doses of most exposures to the general population are likely to be much lower than what would be predicted by the linear model that has been the default in U.S. regulatory risk assessment. The true risk might often be zero.

Agencies that evaluate potential cancer risks to humans are moving to take mechanism and non-linearity into account. The U.S. EPA recently proposed new cancer risk assessment guidelines[37] that emphasize a more flexible approach to risk assessment and call for use of more biological information in the weight-of-evidence evaluation of carcinogenicity for a given chemical and in the dose–response assessment. The proposed changes take into account the issues that we have discussed above. The new EPA guidelines recognize the dose-dependence of many toxicokinetic and metabolic processes, and the importance of understanding cancer mechanisms for a chemical. The guidelines use non-linear approaches to low-dose extrapolation if warranted by mechanistic data and a possible threshold of dose below which effects will not occur. In addition, toxicological results for cancer and non-cancer endpoints could be incorporated together in the risk assessment process.

Also consistent with the results we discussed above are the recent IARC consensus criteria for evaluations of carcinogenicity in rodent studies, which take into account that an agent can cause cancer in laboratory animals through a mechanism that does not operate in humans.[18] The tumors in such cases involve persistent hyperplasia in cell types from which the tumors arise. These include urinary bladder carcinomas associated with certain urinary precipitates, thyroid follicular-cell tumors associated with altered thyroid stimulating hormone (TSH), and cortical tumors of the kidney that arise only in male rats in association with nephropathy that is due to α_2 urinary globulin.

Historically, in U.S. regulatory policy, the 'virtually safe dose', corresponding to a maximum, hypothetical risk of one cancer in a million, has routinely been estimated from results of carcinogenesis bioassays using a linear model, which assumes that there are no unique effects of high doses. To the extent that carcinogenicity in rodent bioassays is due to the effects of high doses for the non-mutagens, and a synergistic effect of cell division at high doses with DNA damage for the mutagens, then this model overestimates risk.[27,38]

We have discussed validity problems associated with the use of the limited data from animal cancer tests for human risk assessment.[39] Standard practice in regulatory risk assessment for a given rodent carcinogen has been to extrapolate from the high doses of rodent bioassays to the low doses of most human exposures by multiplying carcinogenic potency in rodents by human exposure. Strikingly, however, owing to the relatively narrow range of doses in two-year rodent bioassays and the limited range of statistically significant tumor incidence

[37] U.S. Environmental Protection Agency, *Fed. Reg.*, 1996, **61**, 17960.
[38] D. W. Gaylor and L. S. Gold, *Regul. Toxicol. Pharmacol.*, 1998, **28**, 222; http://www.idealibrary.com/links/doi/10.1006/rtph.1998.1258/pdf.
[39] L. Bernstein, L. S. Gold, B. N. Ames, M. C. Pike and D. G. Hoel, *Fundam. Appl. Toxicol.*, 1985, **5**, 79.

rates, the various measures of potency obtained from two-year bioassays, such as the EPA q_1^* value, the TD_{50}, and the lower confidence limit on the TD_{10} (LTD_{10}) are constrained to a relatively narrow range of values about the MTD, in the absence of 100% tumor incidence at the target site, which rarely occurs.[12,34,38-40] For example, the dose usually estimated by regulatory agencies to give one cancer in a million can be approximated simply by using the MTD as a surrogate for carcinogenic potency. The 'virtually safe dose' (VSD) can be approximated from the MTD. Gaylor and Gold[40] used the ratio MTD/TD_{50} and the relationship between q_1^* and TD_{50} to estimate the VSD. The VSD was approximated by the MTD/740,000 for rodent carcinogens tested in the bioassay program of the National Cancer Institute (NCI)/National Toxicology Program (NTP). The MTD/740 000 was within a factor of 10 of the VSD for 96% of carcinogens. This variation is similar to the variation in potency when the same chemical is tested twice in the same strain and sex by the same route: in such near-replicate experiments, potency estimates vary by a factor of four around a median value.[12]

Using the newly proposed benchmark dose of the U.S. EPA carcinogen guidelines, risk estimation is similarly constrained by bioassay design. A simple, quick, and relatively precise determination of the LTD_{10} can be obtained by the MTD divided by seven.[38] Both linear extrapolation and the use of safety or uncertainty factors proportionately reduce a tumor dose in a similar manner. The difference in the regulatory 'safe dose,' if any, for the two approaches depends on the magnitude of uncertainty factors selected. Using the benchmark dose approach of the proposed EPA carcinogen risk assessment guidelines, the dose estimated from the LTD_{10} divided, *e.g.* by a 10 000-fold uncertainty factor, is similar to the dose of an estimated risk of less than 10^{-5} using a linear model. This dose is 10 times higher than the virtually safe dose corresponding to an estimated risk of less than 10^{-6}. Thus, whether the procedure involves a benchmark dose or a linearized model, cancer risk estimation is constrained by bioassay design.

5 The HERP Ranking of Possible Carcinogenic Hazards

Given the lack of epidemiological data to link pesticide residues to human cancer, as well as the limitations of cancer bioassays for estimating risks to humans at low exposure levels, the high positivity rate in bioassays, and the ubiquitous human exposures to naturally occurring chemicals in the normal diet that are rodent carcinogens (Tables 1, 2, and 3), how can bioassay data best be used to evaluate potential carcinogenic hazards to humans? We have emphasized the importance of gaining a broad perspective about the vast number of chemicals to which humans are exposed. A comparison of potential hazards can be helpful in efforts to communicate to the public what might be important cancer prevention factors, when setting research and regulatory priorities, and when selecting chemicals for chronic bioassay, mechanistic, or epidemiologic studies.[4,12] There is a need to identify what might be the important cancer hazards among the ubiquitous exposures to rodent carcinogens in everyday life.

One reasonable strategy for setting priorities is to use a rough index to *compare*

[40] D. W. Gaylor and L. S. Gold, *Regul. Toxicol. Pharmacol.*, 1995, **22**, 57; http://www.idealibrary.com/links/doi/10.1006/rtph.1995.1069/pdf.

and *rank* possible carcinogenic hazards from a wide variety of chemical exposures at levels that humans receive, and then to focus on those that rank highest in possible hazard.[4] Ranking is thus a critical first step. Although one cannot say whether the ranked chemical exposures are likely to be of major or minor importance in human cancer, it is not prudent to focus attention on the possible hazards at the bottom of a ranking if, using the same methodology to identify a hazard, there are numerous common human exposures with much greater possible hazards. Our analyses are based on the HERP index (Human Exposure/Rodent Potency), which indicates what percentage of the rodent carcinogenic potency (TD_{50} in mg/kg/day) a human receives from a given average daily dose for a lifetime of exposure (mg/kg/day). TD_{50} values in our CPDB span a 10 million-fold range across chemicals.[12,13] Human exposures to rodent carcinogens range enormously as well, from historically high workplace exposures in some occupations to very low exposures from residues of synthetic chemicals.

The rank order of possible hazards for a given exposure by the simple HERP index will be similar to a ranking of regulatory 'risk estimates' using a linear model, since they are both proportional to dose. Overall, our analyses have shown that synthetic pesticide residues rank low in possible carcinogenic hazards compared to many common exposures. HERP values for some historically high exposures in the workplace and some pharmaceuticals rank high, and there is an enormous background of naturally occurring rodent carcinogens in typical portions or average consumption of common foods that casts doubt on the relative importance of low-dose exposures to residues of synthetic chemicals such as pesticides. A committee of the National Research Council recently reached similar conclusions about natural *vs.* synthetic chemicals in the diet, and called for further research on natural chemicals[41] (see Section 6 below for further work on natural chemicals).

The HERP ranking in Table 5 is for *average* U.S. exposures to all rodent carcinogens in the Carcinogenic Potency Database for which concentration data and average exposure or consumption data were both available and for which human exposure could be chronic for a lifetime. For pharmaceuticals the doses are recommended doses, and for workplace they are past industry or occupation averages. The 87 exposures in the ranking (Table 5) are ordered by possible carcinogenic hazard (HERP), and natural chemicals in the diet are reported in boldface.

Several HERP values make convenient reference points for interpreting Table 5. The median HERP value is 0.002%, and the background HERP for the average chloroform level in a liter of U.S. tap water is 0.0003%. Chloroform is formed as a by-product of chlorination. A HERP of 0.00001% is approximately equal to a U.S. regulatory VSD risk of 10^{-6}. Using the benchmark dose approach recommended in the new EPA guidelines with the LTD_{10} as the point of departure (POD), linear extrapolation would produce a similar estimate of risk at 10^{-6} and hence a similar HERP value.[38] If information on the carcinogenic mode of action for a chemical supports a non-linear dose–response curve, then

---

[41] National Research Council, *Carcinogens and Anticarcinogens in the Human Diet: A Comparison of Naturally Occurring and Synthetic Substances*, National Academy Press, Washington, 1996.

the EPA guidelines call for a margin of exposure approach with the LTD_{10} as the POD. The reference dose using a safety or uncertainty factor of 1000 (*i.e.* $LD_{10}/1000$) would be equivalent to a HERP value of 0.001%. If the dose–response is judged to be non-linear, then the cancer risk estimate will depend on the number and magnitude of safety factors used in the assessment.

The HERP ranking maximizes possible hazards to synthetic chemicals because it includes historically high exposure values that are now much lower, *e.g.* DDT, saccharin, and some occupational exposures. Additionally, the values for dietary pesticide residues are averages in the *total diet*, whereas for most natural chemicals the exposure amounts are for concentrations of a chemical in an individual food (*i.e.* foods for which data are available on concentration and average U.S. consumption).

Table 5 indicates that many ordinary foods would not pass the regulatory criteria used for synthetic chemicals. For many natural chemicals the HERP values are in the top half of the table, even though natural chemicals are markedly under-represented because so few have been tested in rodent bioassays. We discuss several categories of exposure below and indicate that mechanistic data are available for some chemicals, which suggest that the possible hazard may not be relevant to humans or would be low if non-linearity or a threshold were taken into account in risk assessment.

Occupational Exposures

Occupational and pharmaceutical exposures to some chemicals have been high, and many of the single chemical agents or industrial processes evaluated as human carcinogens have been identified by historically high exposures in the workplace. HERP values rank at the top of Table 5 for chemical exposures in some occupations to ethylene dibromide, 1,3-butadiene, tetrachloroethylene, formaldehyde, acrylonitrile, trichloroethylene, and methylene chloride. When exposures are high, the margin of exposure from the carcinogenic dose in rodents is low. The issue of how much human cancer can be attributed to occupational exposure has been controversial, but a few percent seems a reasonable estimate.[1]

Pharmaceuticals

Some pharmaceuticals that are used chronically are also clustered near the top of the HERP ranking, *e.g.* phenobarbital, clofibrate, and fluvastatin. In Table 3 we reported that half the drugs in the PDR with cancer test data are positive in rodent bioassays.[22] Most drugs, however, are used for only short periods, and the HERP values for the rodent carcinogens would not be comparable to the chronic, long-term administration used in HERP. The HERP values for less than chronic administration at typical doses would produce high HERP values, *e.g.* phenacetin (0.3%), metronidazole (5.6%), and isoniazid (14%).

Herbal supplements have recently developed into a large market in the U.S.; they have not been a focus of carcinogenicity testing. The FDA regulatory requirements for safety and efficacy that are applied to pharmaceuticals do not pertain to herbal supplements under the 1994 Dietary Supplements and Health

Table 5 Ranking possible carcinogenic hazards from average U.S. exposures to rodent carcinogens[a]

Possible hazard: HERP (%)	Average daily US exposure	Human dose of rodent carcinogen	Potency TD$_{50}$ (mg/kg/day)[b]	
			Rats	Mice
140	EDB: production workers (high exposure) (before 1977)	Ethylene dibromide, 150 mg	1.52	(7.45)
17	Clofibrate	Clofibrate, 2 g	169	ND
14	Phenobarbital, 1 sleeping pill	Phenobarbital, 60 mg	(+)	6.09
6.8	1,3-Butadiene: rubber industry workers (1978–86)	1,3-Butadiene, 66.0 mg	(261)	13.9
6.2	**Comfrey-pepsin tablets, 9 daily (no longer recommended)**	**Comfrey root, 2.7 g**	626	ND
6.1	Tetrachloroethylene: dry cleaners with dry-to-dry units (1980–90)	Tetrachloroethylene, 433 mg	101	(126)
4.0	Formaldehyde: production workers (1979)	Formaldehyde, 6.1 mg	2.19	(43.9)
2.4	Acrylonitrile: production workers (1960–1986)	Acrylonitrile, 28.4 mg	16.9	ND
2.2	Trichloroethylene: vapor degreasing (before 1977)	Trichloroethylene, 1.02 g	668	(1580)
2.1	**Beer, 257 g**	**Ethyl alcohol, 13.1 ml**	9110	(−)
1.4	Mobile home air (14 h/day)	Formaldehyde, 2.2 mg	2.19	(43.9)
1.3	**Comfrey-pepsin tablets, 9 daily (no longer recommended)**	**Symphytine, 1.8 mg**	1.91	ND
0.9	Methylene chloride: workers, industry average (1940s–80s)	Methylene chloride, 471 mg	724	(1100)
0.5	**Wine, 28.0 g**	**Ethyl alcohol, 3.36 ml**	9110	(−)
0.5	Dehydroepiandrosterone (DHEA)	DHEA supplement, 25 mg	68.1	ND
0.4	Conventional home air (14 h/day)	Formaldehyde, 598 μg	2.19	(43.9)
0.2	Fluvastatin	Fluvastatin, 20 mg	125	ND
0.1	**Coffee, 13.3 g**	**Caffeic acid, 23.9 mg**	297	(4900)
0.1	**d-Limonene in food**	**d-Limonene, 15.5 mg**	204	(−)
0.04	Lettuce, 14.9 g	Caffeic acid, 7.90 mg	297	(4900)
0.03	**Safrole in spices**	**Safrole, 1.2 mg**	(441)	51.3
0.03	Orange juice, 138 g	d-Limonene, 4.28 mg	204	(−)
0.03	**Comfrey herb tea, 1 cup (1.5 g root) (no longer recommended)**	**Symphytine, 38 μg**	1.91	ND

0.03	Tomato, 88.7 g	Caffeic acid, 5.46 mg	297	(4900)
0.03	Pepper, black, 446 mg	d-Limonene, 3.57 mg	204	(–)
0.02	Coffee, 13.3 g	Catechol, 1.33 mg	88.8	(244)
0.02	Furfural in food	Furfural, 2.72 mg	(683)	197
0.02	Mushroom (*Agaricus bisporus* 2.55 g)	Mixture of hydrazines, *etc.* (whole mushroom)	ND	20 300
0.02	Apple, 32.0 g	Caffeic acid, 3.40 mg	297	(4900)
0.02	Coffee, 13.3 g	Furfural, 2.09 mg	(683)	197
0.01	BHA: daily US avg (1975)	BHA, 4.6 mg	606	(5530)
0.01	Beer (before 1979), 257 g	Dimethylnitrosamine, 726 ng	0.0959	(0.189)
0.008	Aflatoxin: daily US avg (1984–89)	Aflatoxin, 18 ng	0.0032	(+)
0.007	Cinnamon, 21.9 mg	Coumarin, 65.0 μg	13.9	(103)
0.006	Coffee, 13.3 g	Hydroquinone, 333 μg	82.8	(225)
0.005	Saccharin: daily US avg (1977)	Saccharin, 7 mg	2140	(–)
0.005	Carrot, 12.1 g	Aniline, 624 μg	194[c]	(–)
0.004	Potato, 54.9 g	Caffeic acid, 867 μg	297	(4900)
0.004	Celery, 7.95 g	Caffeic acid, 858 μg	297	(4900)
0.004	White bread, 67.6 g	Furfural, 500 μg	(683)	197
0.003	d-Limonene	Food additive, 475 μg	204	(–)
0.003	Nutmeg, 27.4 mg	d-Limonene, 466 μg	204	(–)
0.003	Conventional home air (14 h/day)	Benzene, 155 μg	(169)	77.5
0.002	Coffee, 13.3 g	4-Methylcatechol, 433 μg	248	ND
0.002	Carrot, 12.1 g	Caffeic acid, 374 μg	297	(4900)
0.002	Ethylene thiourea: daily US avg (1990)	Ethylene thiourea, 9.51 μg	7.9	(23.5)
0.002	BHA: daily US avg (1987)	BHA, 700 μg	606	(5530)
0.002	DDT: daily US avg (before 1972 ban)[d]	DDT, 13.8 μg	(84.7)	12.8
0.001	Plum, 2.00 g	Caffeic acid, 276 μg	297	(4900)
0.001	Pear, 3.29 g	Caffeic acid, 240 μg	297	(4900)
0.001	[UDMH: daily US avg (1988)]	[UDMH, 2.82 μg (from Alar)]	(–)	3.96
0.0009	Brown mustard, 68.4 mg	Allyl isothiocyanate, 62.9 μg	96	(–)
0.0008	DDE: daily US avg (before 1972 ban)[e]	DDE, 6.91 μg	(–)	12.5
0.0006	Bacon, 11.5 g	Diethylnitrosamine, 11.5 ng	0.0266	(+)
0.0006	Mushroom (*Agaricus bisporus* 2.55 g)	Glutamyl-p-hydrazinobenzoate, 107 μg	ND	277
0.0005	Bacon, 11.5 g	Dimethylnitrosamine, 34.5 ng	0.0959	(0.189)
0.0004	Bacon, 11.5 g	N-Nitrosopyrrolidine, 196 ng	(0.799)	0.679
0.0004	EDB: Daily US avg (before 1984 ban)[e]	EDB, 420 ng	1.52	(7.45)
0.0004	Tap water, 1 liter (1987–92)	Bromodichloromethane, 13 μg	(72.5)	47.7
0.0004	TCDD: daily US avg (1994)	TCDD, 6.0 pg	0.0000235	(0.000156)
0.0003	Mango, 1.22 g	d-Limonene, 48.8 μg	204	(–)
0.0003	Beer, 257 g	Furfural, 39.9 μg	(683)	197
0.0003	Tap water, 1 liter (1987–92)	Chloroform, 17 μg	(262)	90.3
0.0003	Carbaryl: daily US avg (1990)	Carbaryl, 2.6 μg	14.1	(–)

Table 5 (*cont.*)

Possible hazard: HERP (%)	Average daily US exposure	Human dose of rodent carcinogen	Potency TD$_{50}$ (mg/kg/day)[b] Rats	Mice
0.0002	**Celery, 7.95 g**	**8-Methoxypsoralen, 4.86 µg**	32.4	(−)
0.0002	Toxaphene: daily US avg (1990)[d]	Toxaphene, 595 ng	(−)	5.57
0.00009	**Mushroom (*Agaricus bisporus*, 2.55 g)**	**p-Hydrazinobenzoate, 28 µg**	ND	454[c]
0.00008	PCBs: daily US avg (1984–86)	PCBs, 98 ng	1.74	(9.58)
0.00008	DDE/DDT: daily US avg (1990)[d]	DDE, 659 ng	(−)	12.5
0.00007	**Parsnip, 54.0 mg**	**8-Methoxypsoralen, 1.57 µg**	32.4	(−)
0.00007	**Toast, 67.6 g**	**Urethane, 811 ng**	(41.3)	16.9
0.00006	Hamburger, pan fried, 85 g	PhIP, 176 ng	4.22[c]	(28.6[c])
0.00006	Furfural	Food additive, 7.77 µg	(683)	197
0.00005	**Estragole in spices**	**Estragole, 1.99 µg**	32.4	51.8
0.00005	**Parsley, fresh, 324 mg**	**8-Methoxypsoralen, 1.17 µg**	32.4	(−)
0.00005	Estragole	Food additive, 1.78 µg	ND	51.8
0.00003	Hamburger, pan fried, 85 g	MeIQx, 38.1 ng	1.66	(24.3)
0.00002	Dicofol: daily US avg (1990)	Dicofol, 544 ng	(−)	32.9
0.00001	**Beer, 257 g**	**Urethane, 115 ng**	(41.3)	16.9
0.000006	Hamburger, pan fried, 85 g	IQ, 6.38 ng	1.65[c]	(19.6)
0.000005	Hexachlorobenzene: daily US avg (1990)	Hexachlorobenzene, 14 ng	3.86	(65.1)
0.000001	Lindane: daily US avg (1990)	Lindane, 32 ng	(−)	30.7
0.0000004	PCNB: daily US avg (1990)	PCNB (Quintozene), 19.2 ng	(−)	71.1
0.0000001	Chlorobenzilate: daily US avg (1989)[d]	Chlorobenzilate, 6.4 ng	(−)	93.9
0.00000008	Captan: daily US avg (1990)	Captan, 115 ng	2080	(2110)
0.00000001	Folpet: daily US avg (1990)	Folpet, 12.8 ng	(−)	1550
<0.00000001	Chlorothalonil: daily US avg (1990)	Chlorothalonil, <6.4 ng	828[e]	(−)

[a]**Chemicals that occur naturally in foods are in bold.** *Average daily human exposure*: reasonable daily intakes are used to facilitate comparisons. The calculations assume a daily dose for a lifetime. *Possible hazard*: the human dose of rodent carcinogen is divided by 70 kg to give a mg/kg/day of human exposure, and this dose is given as the percentage of the TD$_{50}$ in the rodent (mg/kg/day) to calculate the Human Exposure/Rodent Potency index (HERP). TD$_{50}$ values used in the HERP calculation are averages calculated by taking the harmonic mean of the TD$_{50}$ values of the positive tests in that species from the Carcinogenic Potency Database. Average TD$_{50}$ values have been calculated separately for rats and mice, and the more potent value is used for calculating possible hazard. References for average food consumption and concentration of chemicals in foods are reported in L. S. Gold, B. N. Ames and T. H. Slone, 'Misconceptions about the causes of cancer', in *Human and Environmental Risk Assessment: Theory and Practice*, ed. D. Paustenbach, Wiley, New York, in press.

[b]ND = no data in CPDB; a number in parentheses indicates a TD$_{50}$ value not used in the HERP calculation because TD$_{50}$ is less potent than in the other species. (−) = negative in cancer tests; (+) = positive cancer test(s) not suitable for calculating a TD$_{50}$.

[c]TD$_{50}$ harmonic mean was estimated for the base chemical from the hydrochloride salt.

[d]No longer contained in any registered pesticide product (US EPA, 1998).

[e]Additional data from the EPA that is not in the CPDB were used to calculate this TD$_{50}$ harmonic mean.

pharmaceutical drugs, the recommended dose is high relative to the rodent carcinogenic dose. Moreover, under DSHEA the safety criteria that have been used for decades by the FDA for food additives that are 'Generally Recognized As Safe' (GRAS) are not applicable to dietary supplements even though supplements are used at higher doses. Comfrey is a medicinal herb whose roots and leaves have been shown to be carcinogenic in rats. The formerly recommended dose of nine daily comfrey-pepsin tablets has a HERP value of 6.2%. Symphytine, a pyrrolizidine alkaloid plant pesticide that is present in comfrey-pepsin tablets and comfrey tea, is a rodent carcinogen; the HERP value for symphytine is 1.3% in the pills and 0.03% in comfrey herb tea. Comfrey pills are no longer widely sold, but are available on the World Wide Web. Comfrey roots and leaves can be bought at health food stores and on the Web and can thus be used for tea, although comfrey is recommended for topical use only in the *PDR for Herbal Medicines*. Poisoning epidemics by pyrrolizidine alkaloids have occurred in the developing world. In the U.S., poisonings, including deaths, have been associated with use of herbal teas containing comfrey.

Dehydroepiandrosterone (DHEA), a natural hormone manufactured as a dietary supplement, has a HERP value of 0.5% for the recommended dose of one daily capsule containing 25 mg DHEA. DHEA is widely taken in hope of delaying aging, and is a fastest-selling product in health food stores. The mechanism of liver carcinogenesis in rats is peroxisome proliferation, like clofibrate.[10] Recent work on the mechanism of peroxisome proliferation in rodents indicates that it is a receptor-mediated response, suggesting a threshold below which tumors are not induced. This mechanism is unlikely to be relevant to humans at any anticipated exposure level. Recent analyses of the molecular basis of peroxisome proliferation conclude that there is an apparent lack of a peroxisome proliferative response in humans.[42] A recent review of clinical, experimental, and epidemiological studies concluded that late promotion of breast cancer in postmenopausal women may be stimulated by prolonged intake of DHEA.[43]

Natural Pesticides

Natural pesticides, because few have been tested, are markedly under-represented in our HERP analysis. Importantly, for each plant food listed, there are about 50 additional untested natural pesticides. Although about 10 000 natural pesticides and their breakdown products occur in the human diet,[14] only 71 have been tested adequately in rodent bioassays (Table 1). Average exposures to many natural-pesticide rodent carcinogens in common foods rank above or close to the median in the HERP table, ranging up to a HERP of 0.1%. These include caffeic acid (in coffee, lettuce, tomato, apple, potato, celery, carrot, plum, and pear), safrole (in spices, and formerly in natural root beer before it was banned), allyl isothiocyanate (mustard), *d*-limonene (mango, orange juice, black pepper), coumarin in cinnamon, and hydroquinone, catechol, and 4-methylcatechol in

[42] N. J. Woodyatt, K. G. Lambe, K. A. Myers, J. D. Tugwood and R. A. Rovert, *Carcinogenesis*, 1999, **20**, 369.
[43] B. A. Stoll, *Eur. J. Clin. Nutr.*, 1999, **53**, 771.

coffee. Some natural pesticides in the commonly eaten mushroom, *Agaricus bisporus*, are rodent carcinogens (glutamyl-*p*-hydrazinobenzoate, *p*-hydrazinobenzoate), and the HERP based on feeding whole mushrooms to mice is 0.02%. For *d*-limonene, no human risk is anticipated because tumors are induced only in male rat kidney tubules with involvement of α_{2u}-globulin nephrotoxicity, which does not appear to be relevant for humans.[18]

Cooking and Preparation of Food

Cooking and preparation of food can also produce chemicals that are rodent carcinogens. Alcoholic beverages are a human carcinogen, and the HERP values in Table 5 for alcohol in average U.S. consumption of beer (2.1%) and wine (0.5%) are high in the ranking. Ethyl alcohol is one of the least potent rodent carcinogens in the CPDB, but the HERP is high because of high concentrations in alcoholic beverages and high U.S. consumption. Another fermentation product, urethane (ethyl carbamate), has a HERP value of 0.00001% for average beer consumption, and 0.00007% for average bread consumption (as toast).

Cooking food is plausible as a contributor to cancer. A wide variety of chemicals are formed during cooking. Rodent carcinogens formed include furfural and similar furans, nitrosamines, polycyclic hydrocarbons, and heterocyclic amines. Furfural, a chemical formed naturally when sugars are heated, is a widespread constituent of food flavor. The HERP value for naturally occurring furfural in average consumption of coffee is 0.02% and in white bread is 0.004%. Furfural is also used as a commercial food additive, and the HERP for total average U.S. consumption as an additive is 0.00006% (Table 5). Nitrosamines are formed from nitrite or nitrogen oxides (NO_x) and amines in food. In bacon the HERP for diethylnitrosamine is 0.0006%, and for dimethylnitrosamine it is 0.0005%.

A variety of mutagenic and carcinogenic heterocyclic amines (HA) are formed when meat, chicken, or fish are cooked, particularly when charred. Compared to other rodent carcinogens, there is strong evidence of carcinogenicity for HA in terms of positivity rates and multiplicity of target sites; however, concordance in target sites between rats and mice for these HA is generally restricted to the liver.[12] Under usual cooking conditions, exposures to HA are in the low ppb range, and the HERP values are low: for HA in pan-fried hamburger, the HERP value for PhIP is 0.00006%, for MeIQx 0.00003%, and for IQ 0.000006%. Carcinogenicity of the three HA in the HERP table, IQ, MeIQx, and PhIP, has been investigated in long-term studies in cynomolgus monkeys. IQ rapidly induced a high incidence of hepatocellular carcinoma. MeIQx, which induced tumors at multiple sites in rats and mice, did not induce tumors in monkeys. The PhIP study is in progress. Metabolism studies indicate the importance of *N*-hydroxylation in the carcinogenic effect of HA in monkeys.[13] IQ is activated via *N*-hydroxylation and forms DNA adducts; the *N*-hydroxylation of IQ appears to be carried out largely by hepatic CYP3A4 and/or CYP2C9/10, and not by CYP1A2, whereas the poor activation of MeIQx appears to be due to a lack of expression of CYP1A2 and an inability of other cytochromes P450, such as CYP3A4 and CYP2C9/10, to *N*-hydroxylate the quinoxalines. PhIP is

114

activated by *N*-hydroxylation in monkeys and forms DNA adducts, suggesting that it would be expected to have a carcinogenic effect.[13]

Synthetic Pesticides

Synthetic pesticides currently in use that are rodent carcinogens in the CPDB and that are quantitatively detected by the FDA Total Diet Study (TDS) as residues in food are all included in Table 5. Many are at the very bottom of the ranking; however, HERP values are about at the median for ethylenethiourea (ETU), UDMH (from Alar) before its discontinuance, and DDT before its ban in the U.S. in 1972. These three synthetic pesticides still rank below the HERP values for many naturally occurring chemicals that are common in the diet. The HERP values in Table 5 are for residue intake by U.S. females age 65 and older, since that group consumes higher amounts of fruits and vegetables than other adult groups, thus maximizing the exposure estimate to pesticide residues. We note that for pesticide residues in the TDS the consumption estimates for children (mg/kg/day in 1986–1991) are within a factor of three of the adult consumption (mg/kg/day).[16]

DDT and similar early pesticides have been a concern because of their unusual lipophilicity and persistence, even though there is no convincing epidemiological evidence of a carcinogenic hazard to humans[44] and although natural pesticides can also bioaccumulate. In a recently completed 24-year study in which DDT was fed to rhesus and cynomolgus monkeys for 11 years, DDT was not evaluated as carcinogenic[13] despite doses that were toxic to both liver and central nervous system. However, the protocol used few animals and dosing was discontinued after 11 years, which may have reduced the sensitivity of the study.[13]

Current U.S. exposure to DDT and its metabolites is in foods of animal origin, and the HERP value is low, 0.00008%. DDT is often viewed as the typically dangerous synthetic pesticide because it concentrates in adipose tissue and persists for years. DDT was the first synthetic pesticide; it eradicated malaria from many parts of the world, including the U.S., and was effective against many vectors of disease such as mosquitoes, tsetse flies, lice, ticks, and fleas. DDT was also lethal to many crop pests, and significantly increased the supply and lowered the cost of fresh, nutritious foods, thus making them accessible to more people. DDT was also of low toxicity to humans. A 1970 National Academy of Sciences report concluded: 'In little more than two decades, DDT has prevented 500 million deaths due to malaria, that would otherwise have been inevitable'.[45] There is no convincing epidemiological evidence, nor is there much toxicological plausibility, that the levels of DDT normally found in the environment or in human tissues are likely to be a significant contributor to human cancer.

DDT was unusual with respect to bioconcentration, and because of its chlorine substituents it takes longer to degrade in nature than most chemicals; however, these are properties of relatively few synthetic chemicals. In addition, many thousands of chlorinated chemicals are produced in nature. Natural pesticides

[44] T. Key and G. Reeves, *Br. Med. J.*, 1994, **308**, 1520.

[45] National Academy of Sciences, *The Life Sciences: Recent Progress and Application to Human Affairs, the World of Biological Research, Requirement for the Future*, Committee on Research in the Life Sciences, Washington, 1970.

can also bioconcentrate if they are fat soluble. Potatoes, for example, naturally contain the fat-soluble neurotoxins solanine and chaconine,[14] which can be detected in the bloodstream of all potato eaters. High levels of these potato neurotoxins have been shown to cause birth defects in rodents.[15]

For ETU the HERP value would be about 10 times lower if the potency value of the EPA were used instead of our TD_{50}; EPA combined rodent results from more than one experiment, including one in which ETU was administered *in utero*, and obtained a weaker potency[46] (the CPDB does not include *in utero* exposures). Additionally, EPA has recently discontinued some uses of fungicides for which ETU is a breakdown product, and exposure levels are therefore currently lower.

In 1984 the EPA banned the agricultural use of ethylene dibromide (EDB), the main fumigant in the U.S., because of the residue levels found in grain, HERP = 0.0004%. This HERP value ranks low, whereas the HERP of 140% for the high exposures to EDB that some workers received in the 1970s is at the top of the ranking.[4] Two other pesticides in Table 5, toxaphene (HERP = 0.0002%) and chlorobenzilate (HERP = 0.0000001%), have been cancelled in the U.S.

Most residues of synthetic pesticides have HERP values below the median. In descending order of HERP these are carbaryl, toxaphene, dicofol, lindane, PCNB, chlorobenzilate, captan, folpet, and chlorothalonil. Some of the lowest HERP values in Table 5 are for the synthetic pesticides, captan, chlorothalonil, and folpet, which were also evaluated in 1987 by the National Research Council (NRC) and were considered by the NRC to have a human cancer risk above 10^{-6}.[47] Why were the EPA risk estimates reported by NRC so high when our HERP values are so low? We have investigated this disparity in cancer risk estimation for pesticide residues in the diet by examining the two components of risk assessment: carcinogenic potency estimates from rodent bioassays and human exposure estimates.[17] We found that potency estimates based on rodent bioassay data are similar whether calculated, as in the NRC report, as the regulatory q_1^* or as the TD_{50} in the CPDB. In contrast, estimates of dietary exposure to residues of synthetic pesticides vary enormously, depending on whether they are based on the Theoretical Maximum Residue Contribution (TMRC) calculated by the EPA *vs.* the average dietary residues measured by the FDA TDS. The EPA's TMRC is the theoretical maximum human exposure anticipated under the most severe field application conditions, which is often a large overestimate compared to the measured residues. For several pesticides, the NRC risk estimate was greater than one in a million whereas the FDA did not detect any residues in the TDS even though the TDS measures residues as low as 1 ppb.[17]

Food Additives

Food additives that are rodent carcinogens can be either naturally occurring (e.g. allyl isothiocyanate, furfural, and alcohol) or synthetic (butylated hydroxyanisole

[46] U.S. Environmental Protection Agency, *Fed. Reg.*, 1992, **57**, 7484.
[47] National Research Council, *Regulating Pesticides in Food: The Delaney Paradox*, National Academy Press, Washington, 1987.

[BHA] and saccharin, Table 5). The highest HERP values for average dietary exposures to synthetic rodent carcinogens in Table 5 are for exposures in the 1970s to BHA (0.01%) and saccharin (0.005%). Both are non-genotoxic rodent carcinogens for which data on the mechanism of carcinogenesis strongly suggest that there would be no risk to humans at the levels found in food.

BHA is a phenolic antioxidant that is Generally Regarded as Safe (GRAS) by the FDA. By 1987, after BHA was shown to be a rodent carcinogen, its use declined six-fold (HERP = 0.002%); this was due to voluntary replacement by other antioxidants, and to the fact that the use of animal fats and oils, in which BHA is primarily used as an antioxidant, has consistently declined in the U.S. The mechanistic and carcinogenicity results on BHA indicate that malignant tumors were induced only at a dose above the MTD at which cell division was increased in the forestomach, which is the only site of tumorigenesis; the proliferation is only at high doses, and is dependent on continuous dosing until late in the experiment.[48] Humans do not have a forestomach. We note that the dose–response for BHA curves sharply upward, but the potency value used in HERP is based on a linear model; if the California EPA potency value (which is based on a linearized multi-stage model) were used in HERP instead of TD_{50}, the HERP values for BHA would be 25 times lower.[49]

Saccharin, which has largely been replaced by other sweeteners, has been shown to induce tumors in rodents by a mechanism that is not relevant to humans. Recently, both NTP and IARC re-evaluated the potential carcinogenic risk of saccharin to humans. NTP delisted saccharin in its *Report on Carcinogens*,[50] and IARC downgraded its evaluation to Group 3, 'not classifiable as to carcinogenicity to humans'.[18] There is convincing evidence that the induction of bladder tumors in rats by sodium saccharin requires a high dose and is related to development of a calcium phosphate-containing precipitate in the urine,[51] which is not relevant to human dietary exposures. In a recently completed 24-year study by NCI, rhesus and cynomolgus monkeys were fed a dose of sodium saccharin that was equivalent to five cans of diet soda daily for 11 years.[13] The average daily dose-rate of sodium saccharin was about 100 times lower than the dose that was carcinogenic to rats.[12,13] There was no carcinogenic effect in monkeys. There was also no effect on the urine or urothelium, no evidence of increased urothelial cell proliferation, or of formation of solid material in the urine.[13] One would not expect to find a carcinogenic effect under the conditions of the monkey study. Additionally, there may be a true species difference because primate urine has a low concentration of protein and is less concentrated (lower osmolality) than rat urine.[13] Human urine is similar to monkey urine in this respect.[26]

For three naturally occurring chemicals that are also produced commercially and used as food additives, average exposure data were available and they are included in Table 5. The HERP values are as follows. For furfural the HERP

48 D. B. Clayson, F. Iverson, E. A. Nera and E. Lok, *Annu. Rev. Pharmacol. Toxicol.*, 1990, **30**, 441.
49 California Environmental Protection Agency, Standards and Criteria Work Group, *California Cancer Potency Factors: Update*, CalEPA, Sacramento, 1994.
50 U.S. National Toxicology Program, *Ninth Report on Carcinogens*, NTP, Research Triangle Park, NC, 2000.
51 S. M. Cohen, *Food Chem. Toxicol.*, 1995, **33**, 715.

value for the natural occurrence is 0.02% compared to 0.00006% for the additive; for *d*-limonene the natural occurrence HERP is 0.1% compared to 0.003% for the additive; and for estragole the HERP is 0.00005% for both the natural occurrence and the additive.

Safrole is the principal component (up to 90%) of oil of sassafras. It was formerly used as the main flavor ingredient in root beer. It is also present in the oils of basil, nutmeg, and mace.[16] The HERP value for average consumption of naturally occurring safrole in spices is 0.03%. In 1960, safrole and safrole-containing sassafras oils were banned from use as food additives in the U.S.[52] Before 1960, for a person consuming a glass of sassafras root beer per day for life, the HERP value would have been 0.2%. Sassafras root can still be purchased in health food stores and can therefore be used to make tea; the recipe is on the World Wide Web.

Mycotoxins

Of the 23 fungal toxins tested for carcinogenicity, 14 are positive (61%) (Table 3). The mutagenic mold toxin, aflatoxin, which is found in moldy peanut and corn products, interacts with chronic hepatitis infection in human liver cancer development.[18] There is a synergistic effect in the human liver between aflatoxin (genotoxic effect) and the hepatitis B virus (cell division effect) in the induction of liver cancer. The HERP value for aflatoxin of 0.008% is based on the rodent potency. If the lower human potency value calculated by FDA from epidemiological data were used instead, the HERP would be about 10-fold lower.[53] Biomarker measurements of aflatoxin in populations in Africa and China, which have high rates of hepatitis B and C viruses and liver cancer, confirm that those populations are chronically exposed to high levels of aflatoxin. Liver cancer is rare in the U.S. Hepatitis viruses can account for half of liver cancer cases among non-Asians and even more among Asians in the U.S.[54]

Ochratoxin A, a potent rodent carcinogen,[12] has been measured in Europe and Canada in agricultural and meat products. An estimated exposure of 1 ng/kg/day would have a HERP value close to the median of Table 5.[10]

Synthetic Contaminants

Polychlorinated biphenyls (PCBs) and tetrachlorodibenzo-*p*-dioxin (TCDD), which have been a concern because of their environmental persistence and carcinogenic potency in rodents, are primarily consumed in foods of animal origin. In the U.S., PCBs are no longer used, but some exposure persists. Consumption in food in the U.S. declined about 20-fold between 1978 and 1986.[55,56] The HERP value for the most recent reporting of the U.S. FDA Total

[52] U.S. Food and Drug Administration, *Fed. Reg.*, 1960, **25**, 12412.
[53] U.S. Food and Drug Administration, *Assessment of Carcinogenic Upper Bound Lifetime Risk from Resulting Aflatoxins in Consumer Peanut and Corn Products. Report of the Quantitative Risk Assessment Committee*, USFDA, Washington, 1993.
[54] M. C. Yu, M. J. Tong, S. Govindarajan and B. E. Henderson, *J. Natl. Cancer Inst.*, 1991, **83**, 1820.
[55] M. J. Gartrell, J. C. Craun, D. S. Podrebarac and E. L. Gunderson, *J. Assoc. Off. Anal. Chem.*, 1986, **69**, 146.
[56] E. L. Gunderson, *J. Assoc. Off. Anal. Chem.*, 1995, **78**, 910.

Diet Study (1984–86) is 0.00008%, towards the bottom of the ranking, and far below many values for naturally occurring chemicals in common foods. It has been reported that some countries may have higher intakes of PCBs than the U.S.[57]

TCDD, the most potent rodent carcinogen, is produced naturally by burning when chloride ion is present, *e.g.* in forest fires or wood burning in homes. The EPA[58] proposes that the source of TCDD is primarily from the atmosphere directly from emissions, *e.g.* incinerators, or indirectly by returning dioxin to the atmosphere.[58] TCDD bioaccumulates through the food chain because of its lipophilicity, and more than 95% of human intake is from animal fats in the diet.[58] Dioxin emissions decreased by 80% from 1987 to 1995, which the EPA attributes to reduced medical and municipal incineration emissions.[58]

The HERP value of 0.0004% for average U.S. intake of TCDD[58] is below the median of the values in Table 5. Recently, the EPA has re-estimated the potency of TCDD based on a body burden dose-metric in humans (rather than intake)[58] and a re-evaluation of tumor data in rodents (which determined two-thirds fewer liver tumors). Using this EPA potency for HERP would put TCDD at the median of HERP values in Table 6, 0.002%.

TCDD exerts many of its harmful effects in experimental animals through binding to the Ah receptor (AhR), and does not have effects in the AhR knockout mouse.[59] A wide variety of natural substances also bind to the AhR (*e.g.* tryptophan oxidation products), and insofar as they have been examined they have similar properties to TCDD,[15] including inhibition of estrogen-induced effects in rodents.[60] For example, a variety of flavones and other plant substances in the diet, and their metabolites, also bind to the AhR, *e.g.* indole-3-carbinol (I3C). I3C is the main breakdown compound of glucobrassicin, a glucosinolate that is present in large amounts in vegetables of the *Brassica* genus, including broccoli, and gives rise to a potent Ah binder, indolecarbazole.[61] The binding affinity (greater for TCDD) and amounts consumed (much greater for dietary compounds) both need to be considered in comparing possible harmful effects. Some studies provide evidence of enhancement of carcinogenicity of I3C. Additionally, both I3C and TCDD, when administered to pregnant rats, resulted in reproductive abnormalities in male offspring.[62] Currently, I3C is in clinical trials for prevention of breast cancer and also is being tested for carcinogenicity by NTP. I3C is marketed as a dietary supplement at recommended doses about 30 times higher than present in the average Western diet.

TCDD has received enormous scientific and regulatory attention, most recently in an ongoing assessment by the U.S. EPA.[58] Some epidemiological studies suggest an association with cancer mortality, but the evidence is not sufficient to establish causality. IARC evaluated the epidemiological evidence

[57] World Health Organization, *Polychlorinated Biphenyls and Terphenyls*, WHO, Geneva, 1993, vol. 140.

[58] U.S. Environmental Protection Agency, *Exposure and Human Health Reassessment of 2,3,7,8-Tetrachlorodibenzo-p-dioxin (TCDD) and Related Compounds. Draft Final*, USEPA, Washington, 2000.

[59] P. M. Fernandez-Salguero, D. M. Hilbert, S. Rudikoff, J. M. Ward and F. J. Gonzalez, *Toxicol. Appl. Pharmacol.*, 1996, **140**, 173.

[60] S. Safe, F. Wang, W. Porter, R. Duan and A. McDougal, *Toxicol. Lett.*, 1998, **102–103**, 343.

[61] C. A. Bradfield and L. F. Bjeldanes, *J. Toxicol. Environ. Health*, 1987, **21**, 311.

[62] C. Wilker, L. Johnson and S. Safe, *Toxicol. Appl. Pharmacol.*, 1996, **141**, 68.

for carcinogenicity of TCDD in humans as limited.[18] The strongest epidemiological evidence was among highly exposed workers for overall cancer mortality. There is a lack of evidence in humans for any specific target organ. Estimated blood levels of TCDD in studies of those highly exposed workers were similar to blood levels in rats in positive cancer bioassays.[18] In contrast, background levels of TCDD in humans are about 100- to 1000-fold lower than in the rat study. The similarity of worker and rodent blood levels and mechanism of the AhR in both humans and rodents were considered by IARC when they evaluated TCDD as a Group 1 carcinogen, in spite of only limited epidemiological evidence. IARC also concluded that 'Evaluation of the relationship between the magnitude of the exposure in experimental systems and the magnitude of the response (*i.e.* dose–response relationships), do not permit conclusions to be drawn on the human health risks from background exposures to 2,3,7,8-TCDD'. The NTP *Report on Carcinogens* recently evaluated TCDD as 'reasonably anticipated to be a human carcinogen', *i.e.* rather than as a known human carcinogen.[50] The EPA draft final report[58] characterized TCDD as a 'human carcinogen' but concluded that 'there is no clear indication of increased disease in the general population attributable to dioxin-like compounds'.[58] Limitations of data or scientific tools were given by the EPA as possible reasons for the lack of observed effects.

In sum, the HERP ranking in Table 5 indicates that when synthetic pesticide residues in the diet are ranked on possible carcinogenic hazard and compared to the ubiquitous exposures to rodent carcinogens, they rank low. Widespread exposures to naturally occurring rodent carcinogens cast doubt on the relevance to human cancer of low-level exposures to synthetic rodent carcinogens. In U.S. regulatory efforts to prevent human cancer, the evaluation of low-level exposures to synthetic chemicals has had a high priority. Our results indicate, however, that a high percentage of both natural and synthetic chemicals are rodent carcinogens at the MTD, that tumor incidence data from rodent bioassays are not adequate to assess low-dose risk, and that there is an imbalance in testing of synthetic chemicals compared to natural chemicals. There is an enormous background of natural chemicals in the diet that rank high in possible hazard, even though so few have been tested in rodent bioassays. In Table 5, 90% of the HERP values are above the level that would approximate a regulatory virtually safe dose of 10^{-6}.

Caution is necessary in drawing conclusions from the occurrence in the diet of natural chemicals that are rodent carcinogens. It is not argued here that these dietary exposures are necessarily of much relevance to human cancer. In fact, epidemiological results indicate that adequate consumption of fruits and vegetables reduces cancer risk at many sites, and that protective factors like intake of vitamins such as folic acid are important, rather than intake of individual rodent carcinogens.

The HERP ranking also indicates the importance of data on the mechanism of carcinogenesis for each chemical. For several chemicals, data have recently been generated which indicate that exposures would not be expected to be a cancer risk to humans at the levels consumed in food (*e.g.* saccharin, BHA, chloroform, *d*-limonene, discussed above). Standard practice in regulatory risk assessment for chemicals that induce tumors in high-dose rodent bioassays has been to extrapolate risk to low dose in humans by multiplying potency by human exposure. Without data on the mechanism of carcinogenesis, however, the true human risk of cancer at low dose is highly uncertain and could be zero.[4,25,33] Adequate risk assessment from animal cancer tests requires more information for a chemical, about pharmacokinetics, mechanism of action, apoptosis, cell division, induction of defense and repair systems, and species differences. More flexible guidelines on risk assessment have recently been proposed by the U.S. EPA. The guidelines recognize the importance of more biological data and call for a more complete hazard evaluation including animal, human, and mechanistic data. In addition, the new guidelines permit the use of non-linear approaches to low-dose extrapolation if warranted by mechanistic data.[37]

6 Ranking Possible Toxic Hazards from Naturally Occurring Chemicals in the Diet

Since naturally occurring chemicals in the diet have not been a focus of cancer research, it seems reasonable to investigate some of them further as possible hazards because they often occur at high concentrations in foods. Only a small proportion of the many chemicals to which humans are exposed will ever be investigated, and there is at least some toxicological plausibility that high dose exposures may be important. Moreover, the proportion positive in rodent cancer tests is similar for natural and synthetic chemicals, about 50% (see Section 3), and the proportion positive among natural plant pesticides is also similar (Table 3).

In order to identify untested dietary chemicals that might be a hazard to humans *if* they were to be identified as rodent carcinogens, we have used an index, HERT, which is analogous to HERP (see Section 5). HERT is the ratio of Human Exposure/Rodent Toxicity in mg/kg/day expressed as a percentage, whereas HERP is the ratio of Human Exposure/Rodent Carcinogenic Potency in mg/kg/day expressed as a percentage. HERT uses readily available LD_{50} values rather than the TD_{50} values from animal cancer tests that are used in HERP. This approach to prioritizing untested chemicals makes assessment of human exposure levels critical at the outset.

The validity of the HERT approach is supported by three analyses. First, we have found that for the exposures to rodent carcinogens for which we have calculated HERP values, the ranking by HERP and HERT are highly correlated (Spearman rank order correlation = 0.89). Second, we have shown that without conducting a 2-year bioassay the regulatory VSD can be approximated by dividing the MTD by 740 000.[40] Since the MTD is not known for all chemicals, and MTD and LD_{50} are both measures of toxicity, acute toxicity (LD_{50}) can reasonably be used as a surrogate for chronic toxicity (MTD). Third, LD_{50} and carcinogenic potency are correlated; therefore, HERT is a reasonable surrogate index for HERP since it simply replaces TD_{50} with LD_{50}.

We have calculated HERT values using LD_{50} values as a measure of toxicity in combination with available data on concentrations of untested natural chemicals in commonly consumed foods and data on average consumption of those foods in the U.S. diet. Literature searches identified the most commonly consumed foods and concentrations of chemicals in those foods. We considered any chemical with available data on rodent LD_{50}, that had a published concentration ≥ 10 ppm in a common food, and for which estimates of average U.S. consumption of that food were available. The natural pesticides among the chemicals in the HERT table are marked with an asterisk. Among the set of 121 HERT values we were able to calculate (Table 6), the HERT ranged across six orders of magnitude. The median HERT value is 0.007%.

It might be reasonable to investigate further the chemicals in the diet that rank highest on the HERT index and that have not been adequately tested in chronic carcinogenicity bioassays in rats and mice. We have nominated to the National Toxicology Program the chemicals with the highest HERT values as candidates for carcinogenicity testing. These include solanine and chaconine, the main alkaloids in potatoes, which are cholinesterase inhibitors that can be detected in the blood of almost all people; chlorogenic acid, a precursor of caffeic acid; and caffeine, for which no standard lifetime study has been conducted in mice. In rats, cancer tests of caffeine have been negative, but one study that was inadequate, because of early mortality, showed an increase in pituitary adenomas.[12,13]

How would the synthetic pesticides that are rodent carcinogens included in the HERP ranking (Table 5) compare to the natural chemicals that have not been tested for carcinogenicity (Table 6), if they were ranked on HERT, *i.e.* using the same measure of a margin of exposure from the LD_{50}? We calculated HERT using LD_{50} values for the synthetic pesticide residues in the HERP table and found that they rank low in HERT compared to the naturally occurring chemicals in Table 6; 88% (107/121) of the HERT values for the natural chemicals in Table 6 rank higher in possible toxic hazard HERT than any HERT value for any synthetic pesticide that is a rodent carcinogen in the HERP table (Table 5). The highest HERT for the synthetic pesticides would be for DDT before the ban in 1970 (0.00004%).

Many interesting natural toxicants are ranked in common foods in the HERT table. Oxalic acid, which is one of the most frequent chemicals in the table, occurs widely in nature. It is usually present as the potassium or calcium salt and also occurs as the free acid. Oxalic acid is reported in many foods in Table 6; the highest contributors to the diet are coffee (HERT = 0.09%), carrot (0.08%), tea (0.02%), chocolate (0.01%), and tomato (0.01%). Excessive consumption of oxalate has been associated with urinary tract calculi and reduced absorption of calcium in humans.[20]

Because of the high concentrations of natural pesticides in spices, we have reported the HERT values for average intake in Table 6, even though spices are not among the foods consumed in the greatest amounts by weight. The highest concentrations of chemicals in Table 6 are found in spices, which tend to have higher concentrations of fewer chemicals (concentrations can be derived from Table 6 by the ratio of the average consumption of the chemical and the average consumption of the food). High concentrations of natural pesticides in spices include those for menthone in peppermint oil (243 000 ppm), γ-terpinene in lemon oil (85 100 ppm), citral in lemon oil (75 000 ppm), piperine in black pepper

(47 100 ppm), geranial in lemon juice (14 400 ppm) and lemon oil (11 300 ppm). Natural pesticides in spices have antibacterial and antifungal activities whose potency varies by spice.[63] A recent study of recipes in 36 countries examined the hypothesis that spices are used to inhibit or kill food-spoilage microorganisms. Results indicate that as the mean annual temperature increases (and therefore so does spoilage potential), there is an increase in the number of spices used and use of the spices that have greatest antimicrobial effectiveness. The authors argue that spices are used to enhance food flavor, but ultimately are continued in use because they help to eliminate pathogens and therefore contribute to health, reproductive success, and longevity.[63]

Cyanogenesis, the ability to release hydrogen cyanide, is widespread in plants, including several foods, of which the most widely eaten are cassava and lima bean. Cassava is eaten widely throughout the tropics, and is a dietary staple for over 300 million people.[64] There are few effective means of removing the cyanogenic glycosides that produce hydrogen cyanide (HCN), and cooking is generally not effective.[64] For lima beans in Table 6, HERT = 0.01%. Ground flax seed, a dietary supplement, contains about 500 ppm hydrogen cyanide glycosides. The HCN in flax seed appear to be inactivated in the digestive tract of primates.[65]

The increasing popularity of herbal supplements in the U.S. raises concerns about possible adverse effects from high doses or drug interactions.[66] Since the recommended doses of herbal supplements are close to the toxic dose, and since about half of natural chemicals are rodent carcinogens in standard animal cancer tests, it is likely that many dietary supplements from plants will be rodent carcinogens that would rank high in possible carcinogenic hazard (HERP) if they were tested for carcinogenicity. Whereas pharmaceuticals are federally regulated for purity, identification, and manufacturing procedures and additionally require evidence of efficacy and safety, dietary supplements are not. We found that several dietary supplements would have ranked high in the HERT table if we had included them by using the recommended dose and the LD_{50} value for the extract: ginger extract (HERT = 0.8%), ginkgo leaf extract (HERT = 0.7%), ginseng extract (HERT = 0.7%), garlic extract (HERT = 0.1%), and valerian extract (HERT = 0.01%). These results argue for greater toxicological testing requirements and regulatory scrutiny of dietary supplements on the grounds that they may be carcinogens in rodents and that, if so, they are likely to rank high in possible carcinogenic hazard.

[63] J. Billing and P. W. Sherman, *Q. Rev. Biol.*, 1998, **73**, 3.

[64] E. Bokanga, A. J. A. Essers, N. Poulter, H. Rosling and O. Tewe (eds.), *International Workshop on Cassava Safety*, International Society for Horticultural Science, Wageningen, Netherlands, 1994, *Acta Hortic.*, vol. 375.

[65] G. Mazza and B. D. Oomah, in *Flaxseed in Human Nutrition*, ed. S. C. Cunnane and L. U. Thompson, AOCS Press, Champaign, IL, 1995, pp. 56–81.

[66] L. S. Gold and T. H. Slone, *Ranking Possible Toxic Hazards of Dietary Supplements Compared to Other Natural and Synthetic Substances. Testimony to the Food and Drug Administration on Dietary Supplements*; http://potency.berkeley.edu/text/fdatestimony.html; Testimony to the Food and Drug Administration on Dietary Supplements, Docket No. 99N-1174, August 16, 1999.

Table 6 Ranking possible toxic hazards to naturally occurring chemicals in food on the HERT index (Human Exposure/Rodent Toxicity)

Possible hazard: HERT (%)	Average daily consumption of food	Average human consumption of chemical	LD_{50} (mg/kg)	
			Rats	Mice
4.3	Coffee, 500 ml (13.3 g)	*Caffeine, 381 mg	(192)	127
0.3	Tea, 60.2 ml (903 mg)	*Caffeine, 29.4 mg	(192)	127
0.3	Potato, 54.9 g	*α-Chaconine, 4.10 mg	(84P)	19P
0.2	Cola, 174 ml	*Caffeine, 20.8 mg	(192)	127
0.1	Coffee, 500 ml	*Chlorogenic acid, 274 mg	4000P	
0.09	Coffee, 500 ml	*Oxalic acid, 25.2 mg	382	
0.09	Black pepper, 446 mg	*Piperine, 21.0 mg	(514)	330
0.08	Carrot, boiled, 12.1 g	*Oxalic acid, 22.7 mg	382	
0.08	Chocolate, 3.34 g	*Theobromine, 48.8 mg	(1265)	837
0.05	Lemon juice, 1.33 ml	*Geranial, 19.2 mg	500	
0.05	Coffee, 500 ml	*Trigonelline, 176 mg	5000	
0.03	Chocolate, 3.34 g	*Caffeine, 2.30 mg	(192)	127
0.02	Tea, 60.2 ml	*Oxalic acid, 6.67 mg	382	
0.02	Isoamyl alcohol: US avg (mostly beer, alcoholic beverages)	Isoamyl alcohol, 18.4 mg	300	
0.01	Beer, 257 ml	Isoamyl alcohol, 13.6 mg	1300	
0.01	Chocolate, 3.34 g	*Oxalic acid, 3.91 mg	382	
0.01	Tomato, 88.7 g	*Oxalic acid, 3.24 mg	382	
0.01	Coffee, 500 ml	2-Furancarboxylic acid, 821 µg		100P
0.01	Lima beans, 559 mg	Hydrogen cyanide, 28.5 µg		3.7
0.01	Potato chips, 5.2 g	*α-Chaconine, 136 µg[a]	(84P)	19P
0.01	Sweet potato, 7.67 g	*Ipomeamarone, 336 µg		50
0.009	Potato, 54.9 g	*α-Solanine, 3.68 mg	590	
0.008	Isobutyl alcohol: US avg	Isobutyl alcohol, 14.1 mg	2460	
0.008	Hexanoic acid: US avg (beer, grapes, wine)	Hexanoic acid, 15.8 mg	3000	(5000)
0.007	Phenethyl alcohol: US avg	Phenethyl alcohol, 8.28 mg	1790	100J
0.007	Carrot, 12.1 g	*Carotatoxin, 460 µg		
0.006	Ethyl acetate: US avg (mostly alcoholic beverages)	Ethyl acetate, 16.5 mg	(5620)	4100
0.005	Celery, 7.95 g	*Oxalic acid, 1.39 mg	382	
0.005	Coffee, 500 ml	*3-Methylcatechol, 203 µg		56V
0.005	Potato, 54.9 g	*Oxalic acid, 1.26 mg	382	
0.004	Beer, 257 ml	Phenethyl alcohol, 5.46 mg	1790	

0.004	Corn, 33.8 g	*Oxalic acid, 1.12 mg	382	317
0.004	Corn, 33.8 g	Methylamine, 906 µg		
0.004	Peppermint oil, 5.48 mg	*Menthone, 1.33 mg	500	800
0.004	White bread, 67.6 g	Propionaldehyde, 2.09 mg	(1410)	
0.004	Beer, 257 ml	Isobutyl alcohol, 6.40 mg	2460	(7300)
0.003	Tomato, 88.7 g	Methyl alcohol, 13.4 mg	5628	
0.003	Wine, 28.0 ml	Isoamyl alcohol, 3.00 mg	1300	300
0.003	Coffee, 500 ml	Pyrogallol, 555 µg		
0.003	Apple, 32.0 g	*Oxalic acid, 704 µg	382	
0.003	Butyl alcohol: US avg (mostly apple, beer)	Butyl alcohol, 1.45 mg	790	317
0.003	Lettuce, 14.9 g	Methylamine, 567 µg	1870	(6800)
0.003	Beer, 257 ml	Propyl alcohol, 3.29 mg	(780)	685
0.002	Banana, 15.7 g	trans-2-Hexenal, 1.19 mg		
0.002	Orange, 10.5 g	*Oxalic acid, 651 µg	382	
0.002	Wine, 28.0 ml	Ethyl lactate, 4.16 mg	(> 5000)	2500
0.002	Tomato, 88.7 g	*p-Coumaric acid, 1.02 mg		657P
0.002	White bread, 67.6 g	Butanal, 3.44 mg	2490	
0.002	Tea, 60.2 ml	*Theobromine, 1.11 mg	(1265)	837
0.002	Apple, 32.0 g	*Epicatechin, 1.28 mg		1000P
0.002	Tomato, 88.7 g	*Tomatine, 621 µg		500
0.002	Beer, 257 ml	Ethyl acetate, 4.42 mg	(5620)	4100
0.002	Lettuce, 14.9 g	*Oxalic acid, 447 µg	382	
0.001	Apple, 32.0 g	*p-Coumaric acid, 573 µg		657P
0.001	Apple, 32.0 g	*Chlorogenic acid, 3.39 mg	4000P	
0.001	Coffee, 500 ml	Maltol, 462 µg	(1410)	550
0.001	Coffee, 500 ml	Nonanoic acid, 188 µg		224V
0.001	5-Methylfurfural: US avg (mostly coffee)	5-Methylfurfural, 1.71 mg	2200	
0.001	β-Pinene: US avg (mostly pepper, lemon oil, nutmeg)	*β-Pinene, 3.28 mg	4700	
0.001	Broccoli, 6.71 g	*Oxalic acid, 268 µg	382	
0.001	Strawberry, 4.38 g	*Oxalic acid, 261 µg	382	
0.0009	Orange juice, 138 ml	Methyl alcohol, 3.48 mg	5628	(7300)
0.0009	α-Pinene: US avg (mostly pepper, nutmeg, lemon oil)	*α-Pinene, 2.25 mg	3700	
0.0009	White bread, 67.6 g	2-Butanone, 1.65 mg	2737	(4050)
0.0008	Coffee, 500 ml	Pyridine, 519 µg	891	(1500)
0.0008	Acetone: US avg (mostly tomato, bread, beer)	Acetone, 1.74 mg	(5800)	3000

125

Table 6 (cont.)

Possible hazard: HERT (%)	Average daily consumption of food	Average human consumption of chemicals	LD$_{50}$ (mg/kg) Rats	Mice
0.0008	Cucumber, pickled, 11.8 g	Dimethylamine, 182 µg	(698)	316
0.0008	Cabbage, raw, 12.9 g	Methylamine, 169 µg		317
0.0007	Tomato, 88.7 g	*Chlorogenic acid, 2.06 mg	4000P	
0.0007	Wine, 28.0 ml	Methyl alcohol, 2.84 ml	5628	(7300)
0.0007	Coffee, 500 ml	2-Methylpyrazine, 894 µg	1800	
0.0007	Coffee, 500 ml	2,6-Dimethylpyrazine, 432 µg	880	
0.0007	Cabbage, raw, green, 12.9 g	*p-Coumaric acid, 303 µg		657P
0.0006	Peach, 9.58 g	*Chlorogenic acid, 1.78 mg	4000P	
0.0006	Black pepper, 446 mg	*3-Carene, 2.00 mg	4800	
0.0006	Cabbage, boiled, 12.9 g	*Oxalic acid, 155 µg	382	
0.0006	Coffee, 500 ml	Butyric acid, 785 µg	2000	
0.0006	Coffee, 500 ml	2,5-Dimethylpyrazine, 399 µg	1020	
0.0005	Coffee, 500 ml	5-Methylfurfural, 798 µg	2200	
0.0005	Grapes, 11 g	*Oxalic acid, 138 µg	382	
0.0005	Grapes, 11 g	*Chlorogenic acid, 1.38 mg	4000P	
0.0005	Black pepper, 446 mg	*β-Pinene, 1.50 mg	4700	
0.0004	Cucumber (raw flesh), 11.8 g	*Oxalic acid, 118 µg	382	
0.0004	Potato chips, 5.2 g	*α-Solanine, 179 µg	590	
0.0004	Coffee, 500 ml	Propanoic acid, 785 µg	2600	
0.0004	Peach, canned, 9.58 g	*Oxalic acid, 115 µg	382	
0.0004	Lettuce, 14.9 g	Benzylamine, 172 µg		600P
0.0004	Lemon juice, 1.33 ml	Octanal, 1.60 mg	5630	
0.0004	α-Phellandrene: US avg (mostly pepper)	*α-Phellandrene, 1.59 mg	5700	
0.0004	White bread, 67.6 g	Hexanal, 1.35 mg	4890	(8292)
0.0004	Black pepper, 446 mg	*α-Pinene, 1.02 mg	3700	
0.0004	Banana, 15.7 g	2-Pentanone, 424 µg	1600	1600
0.0003	Grapes, 11 g	*Epicatechin, 243 µg		1000P
0.0003	Onion, raw, 14.2 g	Dipropyl trisulfide, 189 µg		800
0.0003	Coffee, 500 ml	2-Ethyl-3-methylpyrazine, 186 µg	880	
0.0003	Pear, 3.29 g	*Chlorogenic acid, 823 µg	4000P	
0.0003	Carrot, 12.1 g	*Chlorogenic acid, 780 µg	4000P	
0.0003	Lemon oil, 8 mg	*γ-Terpinene, 681 µg	3650	
0.0003	Lemon oil, 8 mg	*Geranial, 90.4 µg	500	

Possible hazard	Average daily consumption of food	Average human consumption of chemical	LD_{50}	LD_{50}
0.0003	Lemon oil, 8 mg	*β-Pinene, 832 µg	4700	657P
0.0002	Broccoli (raw), 6.71 g	*p-Coumaric acid, 90.6 µg	4960	(6000)
0.0002	Lemon oil, 8 mg	*Citral, 600 µg	16 600	
0.0001	Isoamyl acetate: US avg (mostly beer, banana)	Isoamyl acetate, 1.70 mg		
0.0001	Corn, canned, 33.8 g	Dimethyl sulfide, 324 µg	3300	(3700)
0.0001	Onions, green, cooked, 137 mg	*Oxalic acid, 31.5 µg	382	
0.0001	Coffee, 500 ml	Hexanoic acid, 245 µg	3000	(5000)
0.0001	Pear, 3.29 g	*Epicatechin, 80.9 µg		1000P
0.00007	Nutmeg, 27.4 mg	*Myristicin, 207 µg	4260	
0.00006	Banana, 15.7 g	Methyl alcohol, 236 µg	5628	(7300)
0.00005	Lemon oil, 8 mg	*α-Pinene, 139 µg	3700	
0.00005	Banana, 15.7 g	Isoamyl acetate, 584 µg	16 600	
0.00005	Strawberry, 4.38 g	*Chlorogenic acid, 136 µg	4000P	
0.00004	Black pepper, 446 mg	*α-Phellandrene, 162 µg	5700	
0.00002	Grapefruit juice, 3.29 ml	Methyl alcohol, 95.4 µg	5628	(7300)
0.00002	Lemon oil, 8 mg	*α-Terpinene, 23.2 µg	1680	
0.00001	Lemon oil, 8 mg	*α-Terpineol, 29.6 µg		2830
0.00001	Black pepper, 446 mg	*α-Terpineol, 25.0 µg		2830
0.00001	Garlic, blanched, 53.3 mg	Diallyl disulfide, 2.05 µg	260	
0.00001	Lemon oil, 8 mg	*Terpinolene, 29.6 µg	4390	
0.000008	Garlic, blanched, 53.3 mg	Diallyl trisulfide, 592 ng		100
0.000001	Garlic, blanched, 53.3 mg	Diallyl sulfide, 2.28 µg	2980	

aLD$_{50}$ values are from the Registry of Toxic Effects of Chemical Substances (RTECS). Parentheses indicate the species with the higher (weaker) LD$_{50}$, which is not used in the HERT calculation. *Average daily consumption of food*: the average amount of the food consumed daily per person in the U.S.; when a chemical is listed rather than a food item, the value is the per person average in the total diet. All calculations assume a daily dose for a lifetime. *Possible hazard*: the amount of chemical reported under '*Average human consumption of chemical*' is divided by 70kg to give a mg/kg of human exposure. The HERT is this human dose (mg/kg/day) as a percentage of the rodent LD$_{50}$ (mg/kg). A '*' preceding a chemical name indicates that the chemical is a natural pesticide. References for average food consumption and concentration of chemicals in foods are reported in L. S. Gold, T. H. Slone and B. N. Ames, 'Pesticide residues in food and cancer risk: a critical analysis,' in *Handbook of Pesticide Toxicology*, ed. R. I. Krieger, Academic Press, New York, in press.

bAbbreviations for LD$_{50}$ values: LO = LD$_{LO}$, P = intraperitoneal injection, V = intravenous injection, J = injection (route not specified).

L. Swirsky Gold, T. H. Slone and B. N. Ames

7 Acknowledgements

This work was supported through the University of California, Berkeley, by the National Institute of Environmental Health Sciences Center Grant ESO1896 (B.N.A. and L.S.G.), by support for research in disease prevention from the Dean's Office of the College of Letters and Science (L.S.G. and B.N.A.), by a grant from the National Foundation for Cancer Research (B.N.A.), and from the U.S. Department of Energy DE-AC-03-76SFO0098 through the E.O. Lawrence Berkeley National Laboratory (L.S.G.).

The MAFF Food LINK Research Programmes

CHRISTINA GOODACRE

1 Policy Background to the Food LINK Programmes

The Ministry of Agriculture, Fisheries and Food (MAFF) has responsibility for the UK food chain, working with the devolved administrations in Northern Ireland and Scotland. The Ministry's main aims are to ensure that consumers benefit from competitively priced food, produced to high standards of safety, environmental care and animal welfare, and from a sustainable, efficient food chain. In support of these aims the Ministry funds research. Almost half of the research budget has been targeted on improving sustainability and the economic performance of the agriculture, fisheries and food industries.

The Food and Drink Industry Division in MAFF has specific responsibility for the food, drink and distributive industries. This Division supports strategic research and related technology transfer to advance the underlying technology of the food and drink industries. Funding is largely provided on the basis of matching contributions from industry. The main activity is under the Government-wide LINK scheme for collaborative research. Since the launch of the first food LINK programme in 1989 there has been a succession of food LINK programmes for the food industry (Table 1). Other MAFF policy groups provide funding for LINK programmes for the agriculture, horticulture and fisheries industries (Table 2).

2 The LINK Scheme for Collaborative Research

Scope and Operation

The LINK scheme is the UK Government's principal mechanism for promoting partnerships in pre-competitive research between industry and the research base. It aims to stimulate innovation and wealth creation as well as to improve the quality of life. The scheme offers participants the opportunity to engage with

Issues in Environmental Science and Technology No. 15
Food Safety and Food Quality

C. Goodacre

Table 1 The Food LINK programmes

Launch date	Programme	Government funding	Sponsors[a]
1989	Food Processing Sciences	£11 million	DTI/MAFF
1991	Agro Food Quality	£9 million	MAFF/DTI/BBSRC
1994	Advanced and Hygienic Food Manufacturing	£7 million	MAFF/BBSRC
1998	Eating, Food and Health	£2.5 million	ESRC/BBSRC/MAFF
1999	Food Quality and Safety	£6 million	MAFF/BBSRC

[a]DTI, Department of Trade and Industry; MAFF, Ministry of Agriculture, Fisheries and Food; BBSRC, Biotechnology and Biological Sciences Research Council; ESRC, Economics and Social Sciences Research Council.

Table 2 Other MAFF-sponsored LINK programmes

Launch date	Programme
1992	Technology for Sustainable Farming Systems
1995	Aquaculture
1996	Horticulture
1996	Sustainable Livestock Systems
1997	Sustainable Arable Crop Production

some of the best and most creative minds in the country to tackle scientific and technological challenges. Companies and research organizations throughout the UK can participate. Multinationals can also participate provided that they have significant manufacturing and research operations in the UK, and the benefits of the research are exploited in the UK or the European Economic Area. Small and medium sized enterprises are particularly encouraged to get involved.

LINK covers a wide range of technology and product areas from food and bio-sciences, through engineering to electronics and communications. Each LINK programme supports a number of collaborative projects which typically have a duration of two to three years. Major commercial benefits in terms of new or improved products or processes typically accrue to the industry partners some two years after completion of a LINK project. Networking between projects within programmes is strongly encouraged so that participants can share in the programme's achievements.

Effective programme management is the key to success of any collaborative scheme. Each LINK programme is managed by a Programme Coordinator together with a Programme Management Committee (PMC), whose members are drawn from industry and the research base. The PMC advises on the technical merits including industrial relevance of project proposals, monitors research progress, and encourages commercial exploitation and wider dissemination of research results.

Setting up a New LINK Programme

It is up to individual Government departments and the Research Councils to develop specific LINK programmes that are targeted on areas of science and

130

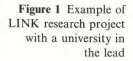

Figure 1 Example of LINK research project with a university in the lead

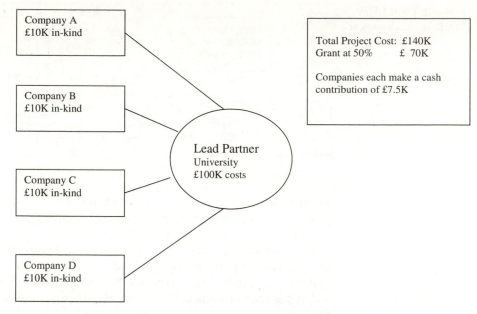

technology with good exploitation potential for the industry sector concerned. This typically involves researching not only industry needs and interests but also the availability of exploitable science and capabilities in the science base. Applications to set up new LINK Programmes are made to the LINK Board, which is a high level committee appointed to oversee the strategic direction and ensure consistent use of the LINK scheme.

LINK programmes typically have a duration of some five years when they are open to new projects and around eight years before the last project is completed. Closed programmes are normally subjected to an evaluation of outputs before any new programme is proposed in a similar technical area. Evaluations examine the specific outputs and value for money from individual research projects. Equally important is to assess the contribution that the programme has made to longer term objectives of raising industry competence levels and coupling the industry more strongly with the science base.

Project Partnerships and Funding

Government sponsors provide up to 50% of the costs of a LINK project, with the balance of funding coming from industry. The industry contribution is usually a mixture of in-kind and cash. Figures 1 and 2 illustrate typical project partnerships in the food LINK programmes.

In the example in Figure 1 the lead partner is a university department which performs most of the research. This is aimed at providing underpinning knowledge for a range of different process operations and products. There are a number of food manufacturing industry partners of equal weight in the project, all aiming to exploit the new knowledge in their own business.

In the example in Figure 2 the lead partner is a food manufacturer who is

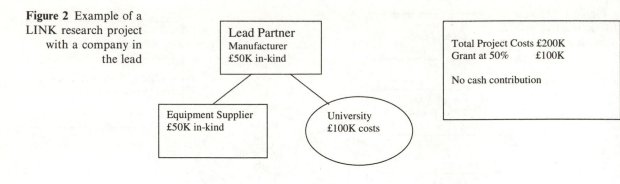

Figure 2 Example of a LINK research project with a company in the lead

working with an equipment manufacturer and a university to research a prototype piece of equipment which will ultimately be developed by the equipment firm for a wider market.

The example in Figure 1 illustrates the general point that cash contributions are required where the industry in-kind contribution is less than the costs of the university research. The balance of in-kind and cash contributions can vary widely between projects and can also vary between industry partners within a project.

An innovative feature of the Advanced and Hygienic Food Manufacturing LINK programme is the introduction of the possibility of funding collaborations where the emphasis is on transfer of technology from non-food industry sectors and demonstrating its applicability in food manufacturing. Up to 20% of programme funds are available for such projects.

A further innovation was introduced by MAFF with the Bridge-LINK scheme which is outside the formal LINK scheme. This provides funds for pre-LINK studies to demonstrate feasibility and take the research up to the point where enough is known to set up a full formal LINK collaboration. Industry takes part in the Bridge-LINK phase and commits to making best efforts to become partners in a follow-on full LINK project.

What is in it for Industry

The LINK scheme provides companies with access to high quality research and leading edge science which can underpin business strategy and innovation. It can bring a range of skills to bear on problems and offers opportunities for networking and sharing ideas with experts in many fields. LINK helps reduce the research investment required by sharing the cost among a number of firms and with Government. This enables industry to pursue highly innovative but risky research objectives and approaches. The contact with the science base provides firms with a 'window' on emerging science and helps them to focus their own confidential research activities.

What is in it for Academic Researchers

LINK research offers opportunities to work with industrial partners on new areas of research, leading to new discoveries and frequently also to opportunities for high quality publications in the scientific literature. LINK is a source of significant extra funding for research and facilities, thus providing a base from which to develop research excellence and contacts and leading to good prospects for gaining further funding from industry as well as public sources. LINK also offers opportunities to work with other disciplines and to gain business and project management skills.

3 Scientific Themes in Ten Years of Food LINK Programmes

Development of Scientific Themes

The development of the first three food LINK programmes, shown in Table 1, was preceded by the deployment of consultants and industrial secondees to help identify priority areas for research along with outline project ideas which had tentative industry support. Thorough background work was thus carried out to ensure that a good case could be made to Ministers in the sponsoring departments as well as to the LINK Board whose approval was also required.

In the early 1990s the UK Government initiated a major technology foresight exercise, partly to guide public sector research. An important output for the food sector was a series of reports which clearly identified key research priorities for the food industry. These reports provided the main inputs to the two most recently launched food LINK programmes (Eating, Food and Health; Food Quality and Safety).

Focus on Food Quality, Food Safety and Manufacturing Efficiency

The scientific themes in successive food LINK programmes have reflected emerging opportunities for particular science and engineering areas to underpin improvements in the broad areas of food quality, food safety and manufacturing efficiency. Table 3 shows the evolution of priority areas in these programmes.

Materials and measurement science have been recurring themes across the programmes, as applied to agricultural raw materials, food ingredients and processed foods. Biotechnological approaches to probing and improving food quality have also been prominent from the start.

The rapidly developing modelling capabilities in the science and engineering base have permeated much of the research, ranging from modelling microbial growth in complex food structures, to flows and temperatures in process equipment, to the distribution of air-borne contaminants in factory environments, to statistical model-based control in food manufacturing.

The links between diet and health, and the factors that affect consumer acceptability, have also been recurring themes, illustrating the breadth of the research interests of the food sector.

4 Scientific Advances and Commercial Benefits through the Food LINK Programmes

The food LINK programmes have encompassed some 120 projects at a total

C. Goodacre

Table 3 Summary of priority areas in successive food LINK programmes

Programme	Description
Food Processing Sciences Launched 1989 Closed 1995	This was the first food LINK programme. The technical priority areas reflected the growing opportunities for: • Predicting food properties from understanding ingredient interactions • Sensor technology • Process modelling • Biotechnology-based sensors and processes
Agro Food Quality Launched 1991 Closed 1997	This programme focused on enhancing food quality with particular emphasis on achieving this through improved raw materials. The programme took a deliberate food chain approach with the first two of the priority areas: • Features of food raw materials that determine quality for processing and for direct consumption • Strategies for enhancing raw material quality • Measurement and understanding of physical and physiological parameters of food choice • Improvement of the nutritional properties of food
Advanced and Hygienic Food Manufacturing Launched 1994	This programme recognized the importance of improved manufacturing technologies, with the following priority areas: • Hygienic processing and control of contaminants • Process simulation in manufacturing • Intelligent control strategies • Production flexibility • Advanced manufacturing systems • Packaging
Eating, Food and Health Launched 1998	This is a wide-ranging programme addressing the link between diet and health and factors affecting food choice
Food Quality and Safety Launched 1999	This programme takes forward a number of the priority areas in the closed FPS and AFQ programmes: • Raw material properties required for processing • Interaction of ingredients • Measurement of quality and safety • Strategies for enhancing food quality and safety

research cost of over £50 million, including the industry contributions. They have drawn in some 200 companies directly into the partnerships and many more through organizations such as the trade associations, the levy bodies and the various intermediate organizations such as Campden and Chorleywood Research Association, Leatherhead Food Research Association, and Brewing Research International. Significant contributions have been made to the science and

134

engineering base underpinning food manufacturing, as indicated in the evaluation described below.

There are numerous success stories, both in terms of the scientific advances made in very challenging areas and in the commercial exploitation of the new knowledge gained. The following project cases, mainly from research underpinning food quality, illustrate some of these challenges and achievements.

Food Flavour Science

LINK AFQ 92: Aroma Release from Foods (1995–1998). This was a very ambitious project set up between the flavour group at Nottingham University and industrial partners from the food, flavour and instrumentation sectors. The overall goal was to develop a real-time method for analysing volatile flavours in the noses of people eating foods and then apply it to determine whether it related better with the sensory perception of foods compared to the conventional flavour analyses. The specification called for a technique that was sensitive (detection limit around ppt or ppb), rapid (people breathe in and out every 5 s), capable of analysing volatiles simultaneously (flavours contain many different components), could tolerate air and water vapour, was safe and caused minimal disturbance to people. Previous attempts had failed to combine all these factors satisfactorily.

The research idea was to use a modified form of classical chemical ionization mass spectrometry in which molecular ions are produced at ambient pressure from sample molecules ionized in a corona discharge. This technique would be inherently more sensitive and had met with some success in measuring air pollutants at very low levels.

The project was high risk and an early milestone was to demonstrate technical feasibility in order to secure approval to continue the research. This was achieved and the consortium moved quickly to secure patent rights on the working prototype for the apparatus to transfer expired air to the API-MS (Atmospheric Pressure Ionization Mass Spectrometer). In broad outline, expired air during eating was sampled through one nostril, then through a heated transfer line and into an ionization source. Dynamic measurements of concentrations in the breath were displayed on screen. With lag times from nose to signal on screen in the order of milliseconds, many people commented on the strange experience of watching what is happening in-nose before their very eyes.

Micromass were one of the consortium partners and were licensed to produce machines for general sale under the MS-Nose name. So far, machines have been sold in the USA and Australasia and the MS-Nose has generated much interest. While the Industry Partners investigated how the new technique could be applied to their fields of business, the academic team started a programme to study flavour release in model systems and in foods to gain a fundamental understanding of the processes involved. Much progress has been made as documented in a series of publications.[1,2] Since the project produced commercially valuable

---

[1] R. S. T. Linforth, I. Baek and A. J. Taylor, *Food Chem.*, 1999, **65**, 77.
[2] I. Baek, R. S. T. Linforth, A. Blake and A. J. Taylor, *Chem. Senses*, 1999, **24**, 155.

results, there has been much interaction between the partners. Collaboration has continued through other LINK projects and through industry funding for a research post in the university.

The technique is now being used in conjunction with analogous techniques for following non-volatile release in-mouth to build a full picture of the flavour signals sensed during eating a food. This information is being analysed against sensory data, obtained from a trained panel. Already, the new techniques are showing clear evidence of interactions between the volatile and non-volatile flavour compounds, which take place at the cognitive level.

The application of API-MS to real food systems is providing fundamental information about volatile release in relation to physiological processes in the mouth as well as the physicochemical processes that control release in foods (fat content, viscosity, etc.). A very practical significance of the technique is that it can drastically shorten the time and effort taken to reformulate the flavour of food products. Overall, the project is considered to have made a step change in flavour research.

Food Polymer Science

Three major collaborative projects to study glassy and rubbery food polymer systems were coordinated by Nottingham University over the period 1988–1996. Twenty-two different companies were involved, varying from the very small to the very large. These studies have broad industry applications and build on ideas championed by Harry Levine and Louise Slade in the US, and Felix Franks in the UK.

Much of the work in the first project focused on investigating the plasticizing effects of water and sugar on the glass transition temperature of the biopolymers amylopectin (the major component of starch), gluten and casein. Differential scanning calorimetry and other techniques were used to provide systematic data on these effects, which are important to understanding, for example, the retrogradation of starch in baked products.

The second important transition which governs the behaviour of polymers is the melting temperature. Again, studies were made of starch–sugar–water systems and it was shown that at low water contents the melting point is dependent only on water content and independent of the sugar content.

Nearly all food products are phase separated at one distance scale or another and the amount of water associated with each phase will be different. It was therefore recognized in the later projects that it was necessary to account for these factors to predict the temperatures of the glass–rubber and crystalline–amorphous transitions. Several approaches were taken to enable the monitoring of unequal partitioning of water in mixed systems based on measuring the effect of hydration of the individual components: starch, proteins and sugars.

In starch–sugar–water systems there is also an issue of the coupling of the motion of molecules of different sizes. The original view that nothing moved below the measured glass transition temperature has gradually been replaced by a recognition that water, at least, is highly mobile. The mobility of intermediate sized molecules within a polymer is being addressed in later research.

As indicated earlier, the research results had important implications for the prediction of starch retrogradation rates in baked products. These rates are governed by the glass transition and melt temperatures and were modelled using theories from synthetic polymer science. This allowed predictions of the effect of temperature in storage and composition factors such as water content, and the type and concentration of added sugars on the kinetics of starch recrystallization.

Using theoretical approaches of this type, combined with NMR, X-ray diffraction and differential scanning calorimetry measurements, research at Nottingham arrived at proposed molecular mechanisms for the role of sugar in the process of starch recrystallization. This work has provided an understanding of the effect of sugar and has explained apparently contradictory observations.

During the course of these projects, ideas have been developed that are fundamental to the understanding of the behaviour of low-water-content foods, such as the loss of crispness of starch-based snack foods on storage. The new knowledge is assisting the shift from a purely empirical approach to storage stability and product formulation, replacing this with something that is more systematic. The Nottingham team does not claim that all the answers can be found in the polymer science approach. Foods are too complicated for that. Nevertheless, the glass transition and molecular mobility have become essential parts of the thinking of leading edge food technologists. The LINK programmes have made a major contribution to the UK science base in this area.

Visual Appearance of Foods

One of the first food projects (LINK FPS6: On-line vision system for multi-product cake line; 1990–1993) investigated a machine vision solution to the inspection of decorated cake products. The target was to provide a real-time inspection system that was also able to 'learn by seeing' so as to avoid extensive reprogramming for every new product. The project partners included the machine vision group at the University of Wales, Cardiff, an equipment supplier and an end user.

The variable features of cake decorations meant that there was no template to go by. It was thus first necessary to research the boundaries of acceptable appearance and to describe this in terms which could be used by a machine vision system to discriminate between different groups of patterns. The acquisition of images on a production line raised interesting issues regarding the effect of lighting conditions, especially where the contrast between the cake decoration and the base was poor.

The research demonstrated clearly the value of combining intelligent knowledge-based systems (KBS) with image processing for inspecting products with variable features. New generic methods for learning attributes of patterns were produced with fast learning characteristics. The complexities of the inspections tasks were, however, such that available hardware was not fast enough to deal with normal line speeds. Further work was therefore needed by the hardware partner in the project to provide the required high-speed, low-cost hardware.

A further image inspection project started in 1995, this time with the objective of 'seeing' beyond the mere surface appearance of a food product (LINK.AFQ59: Disease, damage and sub-surface detection using image inspection; 1995–1999). The objective was to research an automated system for discriminating on-line between potato tubers showing either surface or sub-surface disease or damage. The project partners were the Scottish Agricultural College, the Potato Marketing Board and an equipment supplier.

For surface defects the research showed that the normal reflectance properties of healthy potato tubers are modified in consistent ways in the presence of certain diseases. The research provided information on the discriminant wavebands that could be used in a multi-spectral camera to detect the diseases of interest. Sub-surface bruise detection was achieved using transmitted light. Overall, the project demonstrated technical feasibility of a number of approaches to inspection. These are now subject to further development by the equipment supplier as part of the exploitation plan.

A more recent inspection project is researching colour calibration for food appearance measurement (LINK AFM65: 1998–2001). This research seeks to adapt colour calibration technology used in the textiles industry and apply it to complex food products such as vegetables. Conventional colour measurements for such products ignore the variability across the image and simply provide an average colour measurement. This does not represent what the eye sees.

The new research builds on the substantial progress that has been made in computer-based colour imaging, enabling analysis of CIE (International Lighting Commission) colour information at a pixel level. With this detail, the information received from, for example, a vegetable product will yield many tristimulus values, indicating the variability in colour. This information can be visualized on a calibrated screen in CIE colour space. However, to produce satisfactory visual matches of products it is also necessary to develop a colour appearance model that is capable of predicting the perceived change in appearance for various media and viewing conditions. Such colour appearance models have never before been applied to foods.

Data-based Modelling to Control Food Quality

LINK FPS 106: Advanced Process Supervision (1993–1997). This project aimed to demonstrate how improved analysis of large amounts of process data from a food manufacturing process could provide the basis for enhanced supervision and control. Key targets for control were more consistent quality, a reduction in the effects of raw material variations, and the ability to operate closer to process constraints. The project partners were the University of Newcastle together with a number of food manufacturing companies.

Data-based modelling had already been adopted in petrochemical processing where the academic partner had considerable previous experience. The new project aimed to research the application of similar approaches in a continuous food manufacturing process, using a breakfast cereal line as the test bed. The key aim was to overcome the theoretical process modelling and control challenges posed and to determine a robust implementation procedure.

The first stage in the modelling exercise was to gather representative process data. This proved to be one of the most important stages in model development. It was first necessary to identify the (off-line) variables that quantify quality and then the on-line process variables that were related to quality. Initial process tests were carried out to determine the appropriate logging frequency and accuracy needed. This required statistical analysis of the information from the plant computer system as well as laboratory samples. Data logging and modelling studies followed, first using linear regression approaches (multiple-linear regression, principal component regression and partial least squares) and then moving on to non-linear methods (artificial neural networks). Hybrid modelling, integrating mechanistic and data-based models, was also studied. During the course of the research, novel algorithms and procedures for data analysis were developed.

The quality of the developed models was assessed for accuracy of implementation and subsequent operator use. A PC was linked into the existing data logging system and software produced to allow the models to be implemented on-line and provide operators with real-time advice. Initial conclusions were that the variance of the five output quality measurements had reduced significantly when compared with the process behaviour prior to implementation. The commercial benefits of greater product consistency has provided the incentive for the company to deploy the model-based approach to controlling other lines in their manufacturing operations. A further major deliverable of the project was a software tool box allowing both off-line analysis and on-line implementation of the methods. Discussions are taking place with a software vendor to enable wider access to this toolbox.

Overall, the project proved a theoretical concept in the context of real-time process supervision and has moved the food processing industry to a new and more sophisticated level in the development of process control techniques.

5   Promoting Dissemination and Exploitation

From the outset the Food LINK programmes have promoted networking of the LINK research community and stimulated wider awareness of research outputs through the arrangement of one-day dissemination events. Each research project has thus been presented at least once and often more than once. Some 30 such events have been held, in some cases combining LINK with related research funded by the Research Councils or the European Union. Although LINK partners are bound by the intellectual property agreement entered into at the start of projects, this has in most cases not hindered the dissemination of major research outputs.

LINK project partners are prompted periodically to demonstrate that exploitation is given appropriate consideration as the research results emerge. Final payment of grant is contingent on the submission of a project completion report which, among other things, has to include an exploitation plan. Projects completed since 1998 have been followed up for a period of two years post completion, with further prompts regarding exploitation.

The Teaching Company Scheme is, as a matter of routine, suggested as a possible route by which companies can take the research forward on a one-to-one basis with their university partner. The advantage of this route is that it provides additional manpower for companies whose existing staff may have neither the time nor the expertise to interpret and exploit the new knowledge by applying it to their particular range of products or processes. For many companies the method of choice is to fund a studentship to pursue particular research leads on a one-to-one basis, focusing on company-specific applications.

The programme evaluation report discussed below included a recommendation that the Research Associations should be invited to help stimulate wider exploitation of the Food LINK research outputs. The Research Associations (RAs) were thus invited in 1997 to put forward suggestions on how this could be accomplished. This resulted in the development of a web site for food LINK research, presentations of LINK research to industry panel members of the Campden and Chorleywood Food Research Association, and a programme of visits by a consultant to individual companies to explore opportunities to apply LINK research outputs. These exercises have illustrated the difficulty in achieving significant impact through one-off events. This has stimulated the development and testing of new mechanisms for translating scientific discoveries in the science base into specific applications in companies outside the immediate LINK project partnership. Continuing efforts are being made by MAFF to test innovative approaches, with the aim of widening the uptake by industry of food LINK research outputs.

6 Evaluation of Two Food LINK Programmes

A firm of consultants, Public and Commercial Economic Consultants, was appointed in 1997 to evaluate the FPS and AFQ LINK programmes (see Table 1). Both programmes were closed to new applications but many of the projects, especially in the AFQ programme, were still under way or had not started. The objectives of the exercise were to analyse the extent to which the two programmes had met their specific objectives, and whether they represented value for money to the sponsoring departments. The evaluation was largely survey-based, comprising telephone interviews with 168 partners in a sample of 37 projects (50% of all projects) and face-to-face interviews with 10 projects. In addition, a panel of experts was consulted.

The evaluation concluded that both programmes were effective in relation to most of their objectives. They were associated with important intermediate outcomes and impacts as well as measurable business and economic benefits. Very few of the projects would have gone ahead without the LINK support, even at a reduced scale or at a later date. Overall, it was concluded that the programmes provided value for money.

Many of the participants had been involved on a repeat basis and a very large majority of participants responded that, given the opportunity, they would definitely or probably participate in LINK research again. Both programmes were thought still to have a valid rationale, the inference being that there was a need for a follow-on programme to succeed the FPS and AFQ programmes.

Specific Findings

Key motivating factors for industry participation have been insufficient knowledge and finance to go it alone. The academic partners, on the other hand, have used the programmes effectively to improve their ability to work with industry. They reported significant additional funding resulting from their LINK research, much of it from industry sources.

Most participants in food LINK had already worked collaboratively, although not with the partners in their LINK projects. Collaborations were characterized by active and multi-directional exchange of 'know how' between all the parties, and they tended to work better than was anticipated at the outset.

At the time of the evaluation, nearly two-thirds of participants had made progress towards commercial exploitation of the outcomes of their projects and this frequently involved further R&D. A majority of academic participants had undertaken consulting or contract research on the strength of their LINK involvement.

Three-quarters of participants were able to estimate how large the business and economic benefits were/might be. These benefits take around five years after project start-up to emerge (in the case of industry partners) and three to four years in the case of academic partners.

The view of the panel of experts was that the two programmes had a rationale that was still valid. Even large companies need the stimulus which the programmes provided. The science content was regarded as both relevant and high quality, and in the large majority of projects was of an internationally high standard.

Lessons Learned

The evaluation also pointed to a number of areas for improvement. It was suggested that the dissemination process could be strengthened to stimulate wider exploitation of research outputs, particularly by small firms which were not well represented in the programme.

The main area for improvement was seen by participants to be in speeding up the project approval process and reducing the time to start up after projects are technically approved. Participants also suggested that Project Monitoring Officers (PMOs) appointed by the funding departments can be a valuable technical resource to projects and thus have an important contribution to make. It was noted that some PMOs were more active than others and the suggestion was made that the appointment and briefing of PMOs should aim to achieve a more uniformly high standard. MAFF has taken on board both recommendations for improvement.

7 The Future for Food Research Collaboration

The potential remains for science and engineering research to contribute to enhanced food quality, food safety and manufacturing efficiency. The replacement of current empiricism with a better understanding of food materials, the basis for their functionalities and their interaction with process machinery, will provide a more rational basis for both product and manufacturing process design. The achievement of these goals will require concerted efforts involving supply chains and research scientists and engineers representing a wide range of disciplines. Research collaboration is thus likely to continue but under formats that are adapted to current needs.

Since the early 1990s, when the food LINK programmes were only just starting, many food companies have undergone changes that mean they no longer have the staff resources to participate speculatively in academically driven research. They thus look more critically at proposals that are put to them, expecting more explicit links with prospective business benefits. In light of experience, many firms are also more realistic about the intellectual challenges of interpreting and translating research outputs into specific applications. It appears that competing calls on the time of their most experienced staff means that participation in pre-competitive research is often regarded as a luxury that firms believe they cannot afford.

Other factors detracting from research collaboration are the time and effort required to set up multi-company projects, the lack of total confidentiality, only partial control over the direction of the research, and the long time before benefits begin to accrue. Some of these barriers can be overcome by one-to-one collaborations between firms and the universities and this is the route that some firms are deliberately taking. However, this route is unlikely to enable the more risky and resource-intensive research areas, with potentially higher long-term rewards, to be tackled.

Approaches which government sponsors can take to overcoming barriers to collaboration are to provide small amounts of funding for 'getting to know you' and technical feasibility studies as a preamble to the development of full three-year collaborative projects. This is the rationale behind the recently developed Bridge-LINK scheme in MAFF. Another approach is to stimulate greater industry pull by employing facilitators, experienced at working at the industry/academe interface, to help generate industry-driven research. Most importantly, sponsors must also work with their constituencies to find ways to speed up the process of getting projects started.

During the 1990s there was a steep increase in the provision of funding for food research collaboration from a variety of public funding sources. This included not only national programmes such as LINK and the Teaching Company Scheme but also the Europe-wide Framework programmes, which have seen particularly strong growth. During this period, both industry and the science base have built up a great deal of experience and skill in how to make research collaboration work for both parties.

It may be expected that significant public funds will continue to be provided for collaborative research. This is stimulated by the growing appreciation of the value of drawing together scientists and engineers from different academic disciplines and of drawing together partners in the supply chain. The challenge for funders will be to ensure that programmes are accessible and meet the needs of both industry and the research-base participants. Funders will in addition need to ensure that programmes achieve benefits more widely for the food industry. The challenge for industry and academe is to make effective use of the programmes that are on offer. This means devoting effort on all sides to identify high impact areas and strong business cases for research.

Sensory Assessment of Food Qualities

PETER J. LILLFORD

1 Introduction

What is quality and how should we measure it? Not an easy question to answer, so we will begin by examining what the current practices are throughout the existing food chain from primary producer to consumer.

The primary producer mostly easily measures quality by the value of the sale of the product, which is often related to its appearance and composition. Normally this results in a grading system by which a standard set of parameters are measured such as size, shape, degree of damage, uniformity of colour and some more descriptive compositional variables such as fat content, sugar content, *etc.* These generally agreed 'grades' allow the producer to set targets for his production and husbandry.[1] The manufacturer, on the other hand, will relate his product quality again to composition, but also to sensory characteristics which he believes the consumer values. The retailer has an even more direct measure which can be simply related to the speed at which the product disappears from the shelf and the number of subsequent complaints received on the performance of that product in the consumer's home. Finally, the consumer makes a complicated assessment of quality which relates to a whole host of factors which in many cases are not properly understood. Notice that all these measurements relate to existing products in the food chain, against which quality criteria or quality assurance can be set. A quite different issue is the fundamental understanding of how and why consumers recognize quality and subsequently pay for it. Let us begin at the easier end of this problem, by understanding some of the criteria for the measurement of the quality of existing products. This is usually known as quality control or quality assurance.

2 Quality Control and Quality Assurance

There are certain mandatory measurements that the law requires to be made on

---

[1] United States Department of Agriculture, *U.S. Standards for Grades of Slaughter Cattle*, USDA, Washington, 1997. United States Department of Agriculture, *U.S. Standards for Grades of Fresh Tomatoes, Fed. Reg.*, 1977, **42**, 32 514.

Issues in Environmental Science and Technology No. 15
Food Safety and Food Quality
© The Royal Society of Chemistry, 2001

raw materials and finished products throughout the food chain. First, they must be safe, that is to say free of toxins, bacteria and other hazards which may be encountered when the food is ingested. Next, the law frequently defines compositional requirements to be given such as protein, fat and water content, the presence of additives, whether these are as processing aids, preservatives or texture and flavour improvers, and finally the nutritional contents, amino acids, carbohydrate and calorie level, nutrient balance, *etc.* These legal requirements are usually embodied in national legislation of composition and labelling. The extent to which labelling is required to describe the product itself continually changes and is constantly subject to revision. Innumerable texts have been written on this subject so it will not be discussed further here.

There is, however, another set of optional measurements which in most cases are designed to standardize the product at the point of delivery or the point of manufacture. The law does not require these measurements to be made but frequently producers, whether of primary or manufactured products, use these techniques simply to assure themselves and their customers that the product is of a constant and reliable performance, so that consistency at the point of purchase is assured to the consumer relative to their expectation. Ideally these tests would be on-line or immediately post-line with regard to the manufacturing process, relate exactly with consumer perception and preference, and be cheap. This, at the moment, is quite impossible, although the search for instrumental methods which directly relate to consumer perception is constantly under investigation. In the absence of accurate instrumental methods, human tasters are frequently employed. Note that they are not required to make judgements of their personal liking of a product, merely to assess whether there is consistency in the appearance, taste, texture, *etc.*, of a particular product form. Amongst these experts we can place wine tasters, tea tasters, cheese tasters, *etc.* These individuals, by virtue of the fact that they taste many products, have a developed accuracy in measuring sensory parameters. In quality assurance there is no particular interest in the method by which accuracy is achieved, only that it is reproducible. Since this is a fairly routine task, there is a danger that the expert will be fatigued and not maintain the standards of accuracy.

There are, therefore, other methods which are purely objective and derived from some physical or chemical device which produces measurable quantity by a standard method.[2] Provided the instruments are properly calibrated and maintained there should be no drift in these quality assurance tests. Amongst these tests we note that the methods are usually related to the assumed quality that the consumer perceives. For example, we have 'tenderometers', which are mechanical devices measuring forces in deformation involved in the fracture of a food which are believed to relate to the same sensory parameter. In fact, in quality assurance the exact correlation is not necessary since the objective of quality assurance is to maintain the product with the same objective value of the measurement, rather than to understand the product's actual performance during eating. In quality assurance we are only concerned with the drift in a

[2] American Oil Chemists' Society, *Official Methods & Recommended Practices of the AOCS*, 4th ed., 1990, official method Cd 16-81, official method Cc 16-60.

particular product from its standard performance. It is always assumed that this performance already intrinsically has a valuable quality, otherwise the product would never have been bought in the first place. It is important to recognize that quality assurance measurements made on products are required to detect deviations from an accepted norm in flavour, colour, texture *etc.* They do not have an exact correlation or relationship to true perception by the consumer. There is a somewhat mistaken belief that a series of highly quantifiable and highly correlated feedback quality control tests would be of value. In many cases they would be interesting, but the expense which they would add to any manufacturing process would put a product outside the range of its competitors. Instead, the only requirement for a better commercial solution is to standardize the process and the raw materials and simply measure that the resultant product has not deviated significantly from the norm.

So much for the practice of quality assurance in the production world; we now turn to the attributes which consumers measure when making their own personal quality judgements.

3   Consumers and Quality

Simply by considering our own behaviour, we can recognize that we make an enormous range of judgements when purchasing any product. In the case of foods which are purchased on a regular basis and are relatively cheap (compared to other consumer goods such as cars, televisions, refrigerators, *etc.*), our judgements are rapid and frequently not subject to a process of conscious thought. Instead, we behave in a somewhat scripted fashion.[3] That is to say, we have an in-built set of behaviour patterns which we carry out almost automatically. These judgements have been derived and defined by the simple frequency of the operations we carry out. There appear to be two types of properties we measure when choosing a product. They are intrinsic and extrinsic qualities. The intrinsic properties relate to the product's appearance and our remembrance of comparable performance in terms of flavour, texture, convenience of preparation, stability, *etc.* All these are the properties of the product itself. In addition, however, we superimpose on these qualities our expectations of its performance, our own habits of use and any information we have acquired on its price, its market position and the influence of its brand image through advertizing. We *perceive* these properties so rapidly that we scarcely remember doing it. This perception process is only the first stage of our action. The second, which determines whether we choose to purchase, is determined by our judgement as to whether we like the product for both its intrinsic and extrinsic properties. A successful product in the marketplace fulfils all of these attributes and then becomes part of our accepted product usage. With this in mind, it is not too surprising therefore that the simple question to the average consumer 'Why do you like that?' can result in the answer 'I don't know, I just do!' The task of quality investigation therefore becomes a problem of disentangling the enormous number of rapid subconscious judgements that are

[3] R. P. Abelson, *Am. Psychol.*, 1981, **36**, 715.

Figure 1 Consumer test
questionnaire

Turkey soup sample

Dear Homemaker

Accompanying this questionnaire is a sample of Turkey Soup. Prepare the soup according to the directions within the test period. After using the soup, please indicate how much you like or dislike it. Use the scale to indicate your attitude by checking at the point which best describes your feeling about the soup.

Please write any comments you have regarding the soup in the space provided. A completely frank and honest opinion will help us determine the product that will best suit consumer needs.

Signature: _____   Address: _____

Homemaker

Like extremely |—   Comments:_____

Like very much |—   _____

Like moderately |—   _____

Like slightly |—   _____

Neither like nor dislike |—

Dislike slightly |—

Dislike moderately |—

Dislike very much |—

Dislike extremely |—

made in deciding a product's quality and identifying which of these can be manipulated successfully to produce some improvement.

Use of Consumer Questionnaires

Although many consumers cannot give a detailed description of why they prefer or like a product, they are always prepared to give a view. We can begin by controlling the input variables on the samples provided to the consumer whilst asking them to state their liking. Figure 1 shows a simple and typical questionnaire in which the variables are controlled. Notice that the consumers have no information on the extrinsic variables of brand, price, market position, *etc.*, but are given a sample in which their liking is scored under the well-established hedonic liking scale. The input of any consumer will be influenced therefore by their expectation of the product, their habits of usage and their perception of its intrinsic properties. Provided a reasonably high score for liking is achieved, we have learnt that whatever the properties were and however they were judged, the overall product is acceptable. Notice the consumer is given an opportunity to comment. This is not necessarily required but frequently consumers are happy to do so; if a common statement across a broad sector produces a simple comment, then this provides some outline information on what parameters may be determining their overall liking. The selection of consumers to be interrogated is in the hands of the experimenter, so that by using this simple technique, differences in liking between various demographic groups can be mapped.

This is known as a monadic test (only one sample is used), but similar methods

146

can be made more and more sophisticated. For example, in a Paired Comparison Test, two samples are presented to the consumer who is asked to choose the preferred sample. If the difference in formulation or process is known by the experimenter, then a direct relationship between manufacturing process and preference can be drawn. The robustness of the result obviously depends on the number of consumers sampled, and in comparison tests an analysis referring to the Assured Judgement (*i.e.* the probability that the choice is outside a simple random phenomenon) is given by:

$$\text{Assured judgement} = NP + (0.5 + Z\sqrt{N})P$$

where N = number of subjects, P = probability of chance = $\frac{1}{2}$ and Z = deviate value. Thus for 24 panellists

$$\text{AJ} = 24/2 + \frac{0.5 + 1.96}{2}\sqrt{24} = 17.05$$

Therefore if 18 panellists agree, their judgement can be assured to be outside random chance.

At the next level of sophistication a triangle test can be used in which three samples are presented, one of which is different from the other two. The consumer is requested to identify the odd sample and asked the subsequent question of whether the duplicates are preferred or the odd sample is preferred.

A further extension of the Paired Comparison method is the 'Rank Order Method'. All samples of a set to be compared are presented at the same time. The task is to rank the samples in order from least to most preferred on some predefined dimension. It has the advantage that more than two samples can be evaluated in one test. However, since considerable concentration is required by the consumer, it is more appropriate to carry out these tests in a controlled environment. The significance of the rank ordering can again be subjected to mathematical analysis and frequently the Friedman test is applied.[4]

The above examples relate to consumer liking or preference, which they find easy to score. It is possible to probe consumer perception using similar methods provided that the stimuli presented are very simply different from each other. For example, it is equally possible to carry out paired comparison, triangle and rank order tests with consumers in which the levels of a single attribute are measured such as sweetness, saltiness, toughness, *etc.* Again, if the formulations or processing of the presented samples is carefully controlled, one can obtain direct relations between the product and its perceived attributes.

Trained Panellists

We noted earlier that individuals who regularly taste foods are capable of developing highly sophisticated articulation of what they are sensing, the classic example being the wine connoisseur. Whilst some of these individuals may have highly developed senses of taste and smell, the significant difference between them

---

[4] J.H. Pollard, *Handbook of Numerical & Statistical Techniques*, Cambridge University Press, Cambridge, 1977.

Figure 2 Taste panel parameters for assessing ice cream

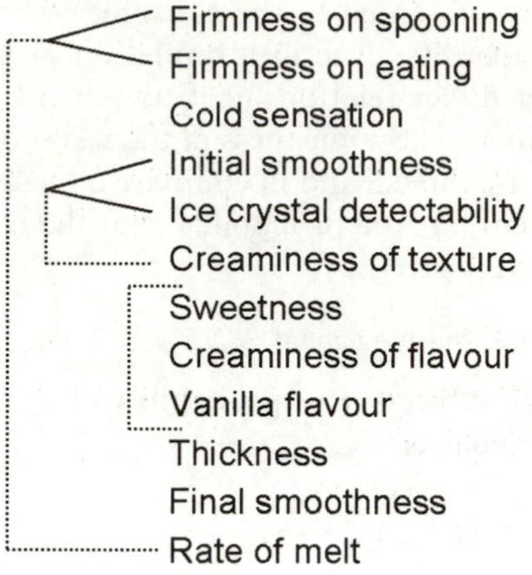

- Firmness on spooning
- Firmness on eating
- Cold sensation
- Initial smoothness
- Ice crystal detectability
- Creaminess of texture
- Sweetness
- Creaminess of flavour
- Vanilla flavour
- Thickness
- Final smoothness
- Rate of melt

Figure 3 Taste panel questionnaire for ice cream

Vanilla ice cream   cc

1.   FIRMNESS TO SPOONING

| 1 | 2 | 3 | 4 | 5 | 6 | 7 | 8 | 9 | 10 | 11 | 12 | 13 | 14 | 15 | 16 | 17 | 18 | 19 | 20 | 20-21 |
Soft   Firm

2.   INITIAL FIRMNESS ON EATING

| 1 | 2 | 3 | 4 | 5 | 6 | 7 | 8 | 9 | 10 | 11 | 12 | 13 | 14 | 15 | 16 | 17 | 18 | 19 | 20 | 22-23 |
Soft   Firm

3.   CHEWINESS

| 1 | 2 | 3 | 4 | 5 | 6 | 7 | 8 | 9 | 10 | 11 | 12 | 13 | 14 | 15 | 16 | 17 | 18 | 19 | 20 | 24-25 |
None   Chewy

4.   COLD SENSATION

| 1 | 2 | 3 | 4 | 5 | 6 | 7 | 8 | 9 | 10 | 11 | 12 | 13 | 14 | 15 | 16 | 17 | 18 | 19 | 20 | 26-27 |
Warm   Cold

5.   INITIAL SMOOTHNESS

etc.....

16.   OVERALL LIKING

| 1 | 2 | 3 | 4 | 5 | 6 | 7 | 8 | 9 | 10 | 11 | 12 | 13 | 14 | 15 | 16 | 17 | 18 | 19 | 20 | 50-51 |
Dislike   Neither like nor dislike   Like

Any comments: _____

and an average consumer is that they are not behaving in a scripted fashion but are applying conscious thought and memory to the sensations they perceive while consuming particular products. Most individuals are capable of carrying out such tasks, but require training in the process of conscious thought and articulation during consumption. An example of this is given in Figure 2, which provides a list of attributes perceived by an average group of consumers when faced with an ice cream, most of whom were not conscious of all the attributes

148

Figure 4 Profile analysis of
a food product

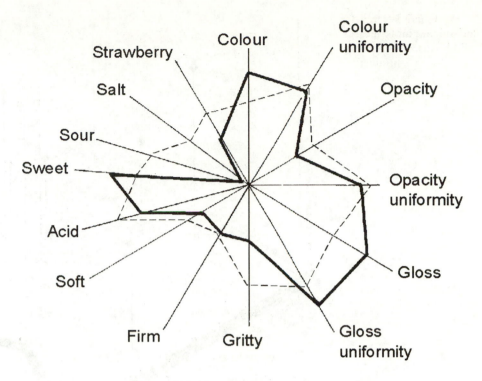

they were detecting until asked to record them. The process of turning a group of individual consumers into a trained panel capable of quantifying all of these attributes requires that the individuals agree on the descriptors used and the level at which the attribute is detected. This is achieved by a training set in which the samples exhibit a variation in all of the descriptors. It is remarkable how quickly a consensus can be reached by individuals on both the descriptor of the attribute and the levels to which it is absent or present in a range of samples. When acceptable agreement has been reached, the panel can be considered 'trained'. It is now possible to present novel samples to this group, who will produce a score on each of the individual parameters assessed. There has been considerable experimentation in the best scale on which panellists should be asked to score,[5] but a simple linear scale from 0 to 10 relative to the training samples is frequently adequate. An example of a typical panellist questionnaire is given in Figure 3 and a dozen or more individual parameters can be scored on any one sample.

Presentation of Trained Panel Data

A trained panel can be considered as a measuring device capable of producing quantitative scores on a significant number of attributes from any presented range of products of similar type. Each sample presented can be described in *n*-dimensional space, where *n* is the number of individual parameters scored for

---

[5] H. L. Meiselman, in *Measurement of Food Preferences*, ed. H. J. H. MacFie and D. M. H. Thomson, Blackie, London, 1994, p. 1.

P. J. Lillford

Figure 5 Three-dimensional plot of meat and textured vegetable products

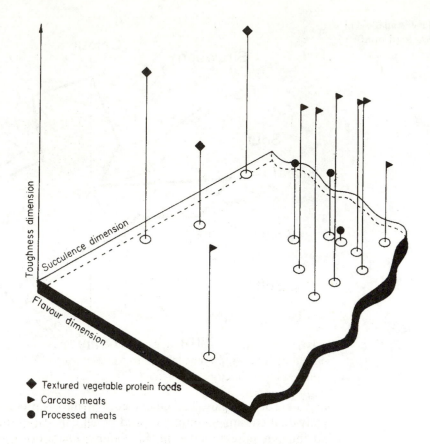

◆ Textured vegetable protein foods
▶ Carcass meats
● Processed meats

each sample. These parameter scores can be considered a fingerprint of the product properties by which samples can be compared (Figure 4). For more than two or three products, however, this comparison becomes difficult and a simplified representation is often required. This means that the dimensions must be reduced to a handleable or visualizable set. A convenient method to do this is Principal Component Analysis, the details of which are described in Horsfield and Taylor.[6] This is particularly useful if the Principal Components can be reduced to a set of three, when a three-dimensional map can be produced. In Figure 5 the main perceived differences in the properties of a set of meat and analogues are presented. The reduced dimensions are Toughness, Succulence and Flavour, each of which contain a primary set of attributes, weighted by their significance in discriminating between the present sample. In this particular example the same set of samples were presented to untrained consumers who were requested to give their liking scores. Their preference can be overlaid on the perception map, providing an indication of those parameters or parameter sets which are most important in determining consumer liking.

[6] S. Horsfield and L. J. Taylor, *J. Sci. Food Agric.*, 1976, **27**, 1044.

150

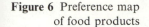

Figure 6 Preference map
of food products

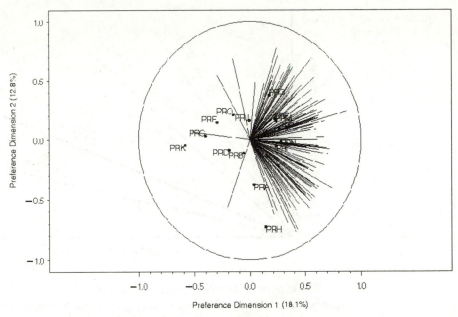

There are many alternative methods of presenting perception data and matching it with preference (or liking). An example of a simpler preference map, represented in circular coordinates, is given in Figure 6. Thus we now have methods which allow a combination of trained panellists and untrained consumers to provide quantitated data relating the difference in products, first to their perceived attributes and second to whether these attributes are desirable. Whilst these provide very valuable correlations between product, formulation, processing and consumer response, they are only correlated and still cannot be related to the causal reasons for perceived differences. To go further we must examine the physical and physiological processes involved in eating and drinking, *i.e.* we must obtain a direct connection between the physical or chemical stimulus and the sensory response.

4   Stimulus Response Measurement

The relationship between a physical or chemical stimulus and the response of a human being is rarely linearly related. Figure 7 shows the result of plotting (on log–log scale) the relation between a physical stimulus such as an electric voltage, weight, sound or light, to the response recorded by a human being. In general the relation is described by Stevens' Law:[7]

$$R = KS^n$$

or

$$\log R = \log K + n \log S$$

where R is the response, S is the Stimulus, n an exponent and K is a constant.

[7] S. S. Stevens, *Science*, 1958, **127**, 383.

151

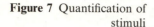

Figure 7 Quantification of
stimuli

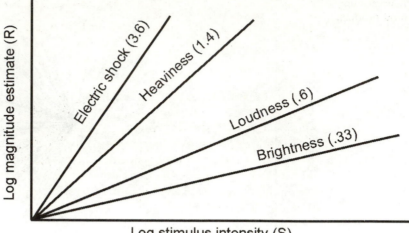

The exponent *n* can be either less than one (*i.e.* the human is less than linearly responsive to the stimulus) or much greater than one, where the human is more sensitive and more reactive as the size of the stimulus increases. It has been argued that this is a simple protection mechanism which we have evolved to be highly sensitive to those external stimuli which may damage us. Notice that these results were obtained by using external stimuli which are easily controlled and measuring the subsequent human response. One of the problems in studying the sensory properties of foods is that whilst we can quantify the response (by using trained panellists), we are frequently unsure of the stimulus causing it, since the perception process takes place in the closed vessel of the mouth where a whole host of mechanoreceptors, chemoreceptors, *etc.*, are distributed. We will begin by examining a simpler case where the quality of the food is perceived external to the mouth, where stimuli are more obviously presented.

Colour and Appearance

These topics have been recently and comprehensively reviewed by Hutchings.[8] Some of the more significant issues are mentioned here.

Colour. We use our colour perception to estimate degrees of ripeness (*e.g.* tomatoes, bananas), extent of cooking (meat, cereal products) and even anticipated flavour strength (tea, fruit juices). Considerable progress in representing colour vision in terms of the three primary colours red, green and blue has been achieved, so that colour matching can be performed by illuminating with only three wavelengths.[9] The colour quality is given by the proportions of each illuminant. To separate this quality from the overall intensity, the colour value is normalized by dividing by the total intensities of the three wavelengths. Thus the colour of a sample (C) is given by

[8] J. B. Hutchings, *Food Colour & Appearance*, 2nd ed., Aspen, Gaithersburg, 1999.
[9] W. D. Wright, *The Measurement of Colour*, 4th ed., Hilger & Watts, London, 1969.

$$C = r[R] + g[G] + b[B]$$

where r, g and b, the chromaticity coordinates, are:

$$r = \frac{R}{R + G + B}; \; g = \frac{G}{R + G + B}; \; b = \frac{B}{R + G + B}$$

This is the basis of Tristimulus Analysis.

Gloss. This is the surface property by which a material appears shiny or lustrous and, in sensory assessment, panellists use these terms. The obverse, however, can be described as dull, grainy or crystalline, indicating the size of surface roughness which can be detected. Clearly the stimulus derives from specular reflectance, so that goniophotometric equipment can be used to quantify and correlate with human perception.

A gloss factor (GF) has been defined as

$$GF = (I_s - I_d)/W_{1/2}$$

where I_s is the peak height at the specular reflectance angle, I_d is the diffuse reflectance intensity and $W_{1/2}$ is the specular peak width at half-height.

For many real samples, having curved surfaces, the observer makes measurements simultaneously over the array of reflectance angles. Not surprisingly, therefore, the simple measurement of GF is often not sufficient to describe stimuli completely.

Translucency. This is the property by which light penetrates and disperses into and through a material. Light scattering and absorbing entities in the structure contribute to the perceived effect. Panellists are usually asked to score specimens from 'clear' to 'opaque', translucent falling in the mid-range of this scale. Some success in relating the panellist scores to the Kubelka–Munk scattering coefficient for model emulsions has been achieved.[10]

Uniformity. Rarely are food products uniform in any of the above parameters. The observer notes the spatial differences in all these properties, which relates to heterogeneity within food materials or between them. Thus, marbling is a difference in colour and translucency across the surface of raw meat, bruising is a colour and gloss difference at and below the surface of fruit and vegetables, and the quality of a plate of peas or sweet corn relates to the sum of all these visual observations exhibited by the population presented to the observer. This is essentially a form of pattern recognition, and the observer compares all the visual information with some 'stored images' of previous experiences. To attempt to understand the complexity of stimuli being analysed by the observer, we need the equivalent optical resolution and computing power available in the human eye and brain. This is now becoming available, as rapid, digital data capture, image processing and reconstruction become available. The necessary physics to create Virtual Reality will be vital in understanding how patterns become interpreted into quality judgements.

[10] R.J. Birkett, in *Proc. 5th Congr. Int. Colour Assoc.*, Monte Carlo, 1985, vol. 1.

Figure 8 Chewed
Longissimus dorsi muscle

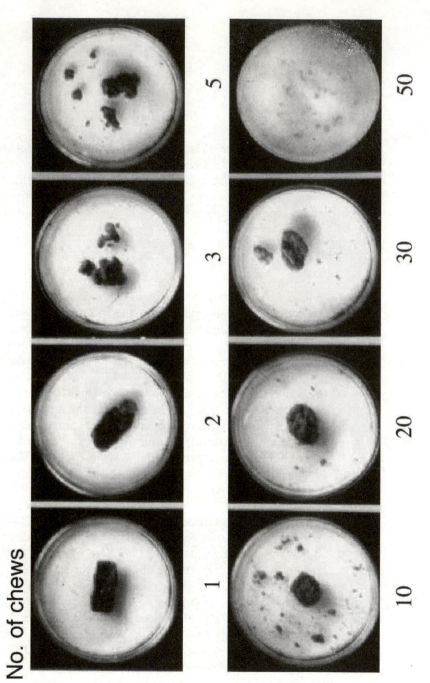

No. of chews

154

Figure 9 Chewed cracker biscuit

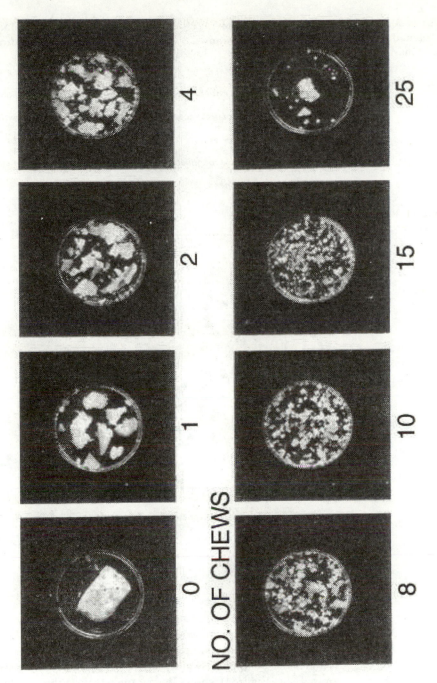

NO. OF CHEWS

Figure 10 Sequence of breakdown of food during mastication

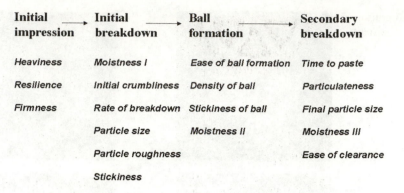

Initial impression	→	Initial breakdown	→	Ball formation	→	Secondary breakdown
Heaviness		Moistness I		Ease of ball formation		Time to paste
Resilience		Initial crumbliness		Density of ball		Particulateness
Firmness		Rate of breakdown		Stickiness of ball		Final particle size
		Particle size		Moistness II		Moistness III
		Particle roughness				Ease of clearance
		Stickiness				

Figure 11 Relation between panel flakiness scores and tensile break strain

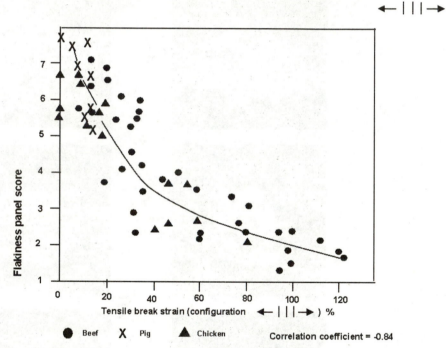

Texture

We begin by qualitative observation of the breakdown pathway of obviously different foods. Figures 8 and 9 show the sequential breakdown of whole meat and a cracker biscuit. Notice that a similar process of fracture and reassembly occurs in these samples. Swallowing is associated with the reassembly process. By comparison, Figure 10 shows the parameters described while eating, but ordered in the sequence in which they occur. We see that physical properties of an initial structure are recorded together with subsequent size and shape of particles, liquid adsorption and release and assembly to a swallowable bolus, and the clearance of material from the palate. Thus texture perception is clearly the measurement of the whole comminution process and we cannot therefore expect the stimuli to be

Figure 12 The mouth process model

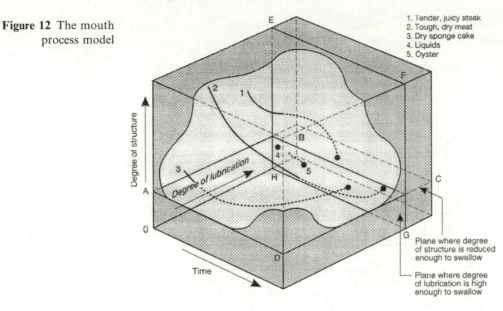

1. Tender, juicy steak
2. Tough, dry meat
3. Dry sponge cake
4. Liquids
5. Oyster

Plane where degree of structure is reduced enough to swallow

Plane where degree of lubrication is high enough to swallow

understood and quantified by simple measurements of the intact food piece. Only the initially perceived attributes are likely to be stimulated from the mechanical properties of the intact food. When reasonable physical tests on the material are related to panellist sensory response, we obtain encouraging correlations. For example, in Figure 11 we show the tensile break strain when cooked meat is pulled at right angles to the fibre direction, compared with the perceived flakiness score. The whole meat kingdom appears to lie on a smooth curve. Note that this is not a simple linear response but shows a relationship more appropriate to Stevens' Law. Conceptually, we can describe texture perception as a process measuring the critical dimensions of 'structure' and 'degree of lubrication' as a function of chewing time.[11] This is shown schematically in Figure 12. Five typical foods are marked on this figure which are (1) a tender juicy steak—structure reduction is mechanically easy and the material also contains sufficient water to lubricate its breakdown; (2) tough, dry meat—more extensive structure reduction is requires lubrication is drawn from saliva and the process takes longer; (3) dry sponge cake—little mechanical work has to be done but extensive saliva mixing is required before swallowing; (4) oyster—the material is close to the swallowable state provided the overall size of the piece is appropriate; (5) liquids, immediately swallowable. Some further evidence for this process has been obtained by asking panellists to score lubrication in the mouth as a function of chewing time. Figure 13 shows three foods. First an orange, from which juices are expressed rapidly on chewing. Second an apple, which releases liquids progressively as cells are ruptured during chewing. Third a dry biscuit; on initial chewing the dry particles require lubrication by saliva. Referring to Figure 12, as well as being capable of perceiving the entire breakdown process, it would appear that our overall

[11] J. B. Hutchings and P. J. Lillford, *J. Texture Studies*, 1998, **19**, 103.

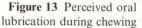

Figure 13 Perceived oral lubrication during chewing

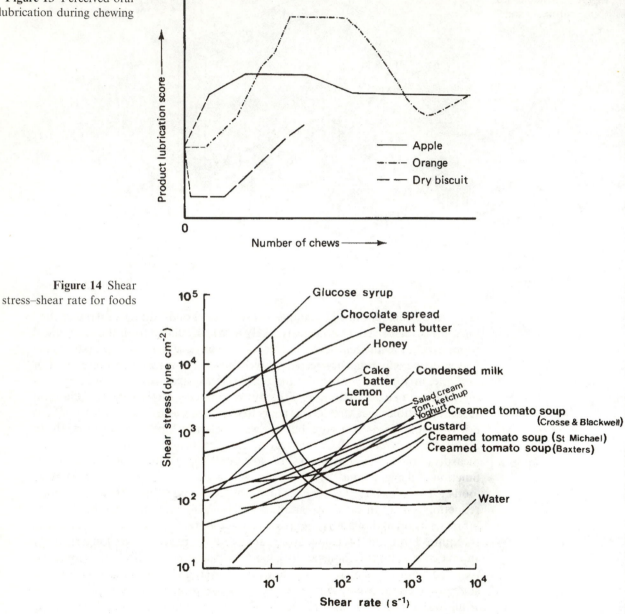

Figure 14 Shear stress–shear rate for foods

preference within a class of foods may relate to specimens which reach the swallowable state in the shortest time.

We can also detect differences in the texture of liquid foods which are more or less immediately swallowable. Shama and Sherman[12] undertook a study of liquid

[12] F. Shama and P. Sherman, *J. Texture Studies*, 1973, **4**, 111.

Figure 15 Taste bud distribution on the tongue and shape of papillae (Reproduced by permission from *Lecture Notes on Human Physiology*, ed. J.J. Bray *et al.*, Blackwell, Oxford, 1989)

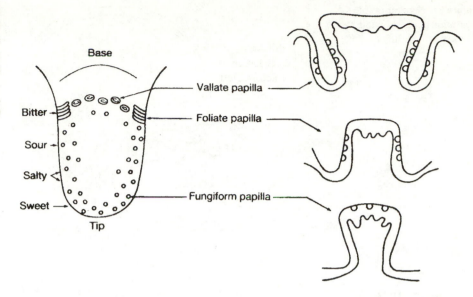

foods of varying viscosities and correlated their actual viscosity with in-mouth perception. Their results are shown in Figure 14, where the shear stress at a given shear rate was measured for each of the foods and panellists were simultaneously asked to score their perception of in-mouth viscosity. The parallel curved lines in Figure 14 show the sensory values. It appears therefore that, for low viscosity liquids, we measure something approaching the shear rate (how fast a sample flows). At higher viscosity the material needs to be pushed between the tongue and the soft palate and so the shear stress becomes the dominant stimulus relating to the perceived properties.

Flavour

It is important to recognize that the sensory property normally described as flavour consists of two elements; these are taste, detected by chemoreceptors in the tongue and soft palate, and odour, detected in the nose by the olfactory receptors. The perception process must therefore be stimulated by the arrival of particular chemical moieties at the site of particular receptors.

Taste.   We know something of the location and type of taste buds in the mouth. Figure 15 shows taste bud distribution and something of their physiological form. They are not evenly distributed on the tongue's surface. It is generally believed that bitter receptors are placed further back in the tongue, that sour receptors are dominantly along the side together with salty receptors, and that sweet-taste receptors are predominantly at the front tip of the tongue. These receptors can be stimulated with different molecular species and it is interesting to note that sweet-tasting molecules such as sugar, aspartame, cyclamate and monellin have

Figure 16 Schematic of human head showing location of olfactory receptors

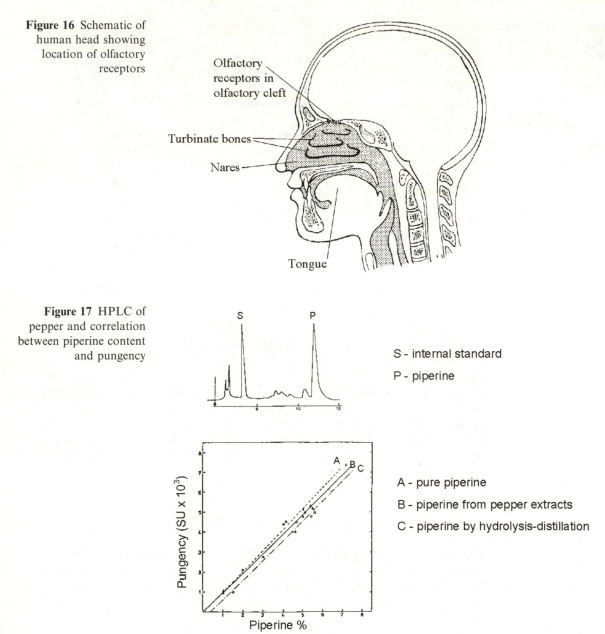

Olfactory receptors in olfactory cleft

Turbinate bones

Nares

Tongue

Figure 17 HPLC of pepper and correlation between piperine content and pungency

S - internal standard

P - piperine

A - pure piperine

B - piperine from pepper extracts

C - piperine by hydrolysis-distillation

Pungency (SU x 10³)

Piperine %

at first sight remarkably different molecular structures, yet all produce a sweet taste sensation in the mouth. Whilst all are sweet, the nature and duration of the sweet taste is very different for each of these molecules, which presumably relates to their binding constant with the chemoreceptors. In most cases the degree of sweetness is not simply related to the concentration of the active species.[13]

---

[13] D. E. Walters and G. Roy, *Flavor–Food Interactions*, ed. R. J. McGorrin and J. V. Leland, ACS Symp. Series 633, American Chemical Society, Washington, 1996.

Figure 18 Effect of fat reduction on level of flavour required to give 40% headspace response

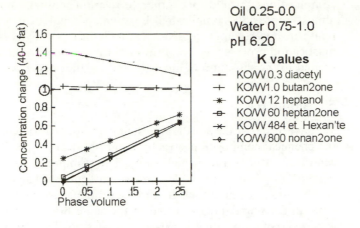

Oil 0.25-0.0
Water 0.75-1.0
pH 6.20

K values

- KO/W 0.3 diacetyl
- KO/W 1.0 butan2one
- KO/W 12 heptanol
- KO/W 60 heptan2one
- KO/W 484 et. Hexan'te
- KO/W 800 nonan2one

Aroma. The detection of odours and aromas takes place in the olfactory cleft in which are positioned thousands of olfactory receptors (Figure 16). The volatiles released from a food or drink enter through the nose or pass round from the back of the throat. The stimulus for odour perception must therefore be the molecular content of the volatile material in space above the food itself. The composition of volatile fraction is easily measured by techniques such as high performance liquid chromatography and in Figure 17 the correlation between sensorily perceived pungency and the piperine content of peppers is shown. In this example the sensory response is directly related to the piperine content. This cannot be a continuous curve since, at high levels of any aroma, saturation can be reached, and no further increase in odour intensity is sensed. The high sensitivity of the chemical detection of volatiles aids flavour analysis and so the production of volatiles by oxidation can be shown to correlate closely with the off-flavour and unacceptability of various soya raw materials. Major complication enters, however, because the detected intensity of odours is not simply related to their concentration in any mixture; some molecular species are much more odour intense than others. During chewing, volatiles will be rapidly sampled as breathing takes place, sweeping the contents of the vapour phase onto the olfactory epithelia. However, the composition of the gaseous phase is not simply related to the composition of the whole food. First, the complex flavour consists of volatile molecules which have differential solubility in aqueous and non-aqueous phases in the food. Therefore, even if equilibrium were reached in the buccal cavity, the concentration in the head space would be dependent on the partition coefficient between individual molecules and their preferred solvent. Figure 18 shows the effect of reducing fat on the levels of independent molecules which is required to give an equivalent flavour to a product containing 40% fat.

There is one further complication which must be considered. The process of mastication also changes the surface structure of all foods and in the case of some fat-continuous emulsions can result in complete inversion to a water-continuous state. This process will be time dependent in the mouth during the chewing process, and this time dependence of structural change will also influence the dynamics of flavour release and therefore the sensory response. With all these

P. J. Lillford

complications it would appear that the relation between stimulus and response is almost impossibly complicated. Fortunately, analytical techniques in the gas phase are now so rapid and sensitive that if the nasal cavity is sampled during the mastication process the composition of the volatiles in the nose can be directly related to the sensation of intensity and balance perceived by the subject.

5 Summary

We have seen that, despite the almost unconscious effort of the average consumer, the chemical composition, structure and physical properties of most foods are sampled routinely and at high speed. The stimuli are converted by our neural network to produce a complex time-dependent set of signals to the brain. This combination of signals and its time-dependence is almost certainly compared with an existing database stored in the same brain which relates to previous experience of the product or expectation of its profile of properties. A judgement of quality appears to be made relative to some standard which the individual has already recognized. We have no physical measurement devices capable of carrying out the same number of measurements and signal integration at equivalent speed. It is not surprising, therefore, that one of the most reliable forms of perception measurement and subsequent quality judgement remains a human being.

However, as we tease apart the significance of chemical composition, physical structure and stability of foods, and our measurement devices are improved to match the sensitivity of and rates of detection which our biological senses provide, we begin to understand both the perception and stimuli and their integration into what everybody regards as 'quality'.

Subject Index